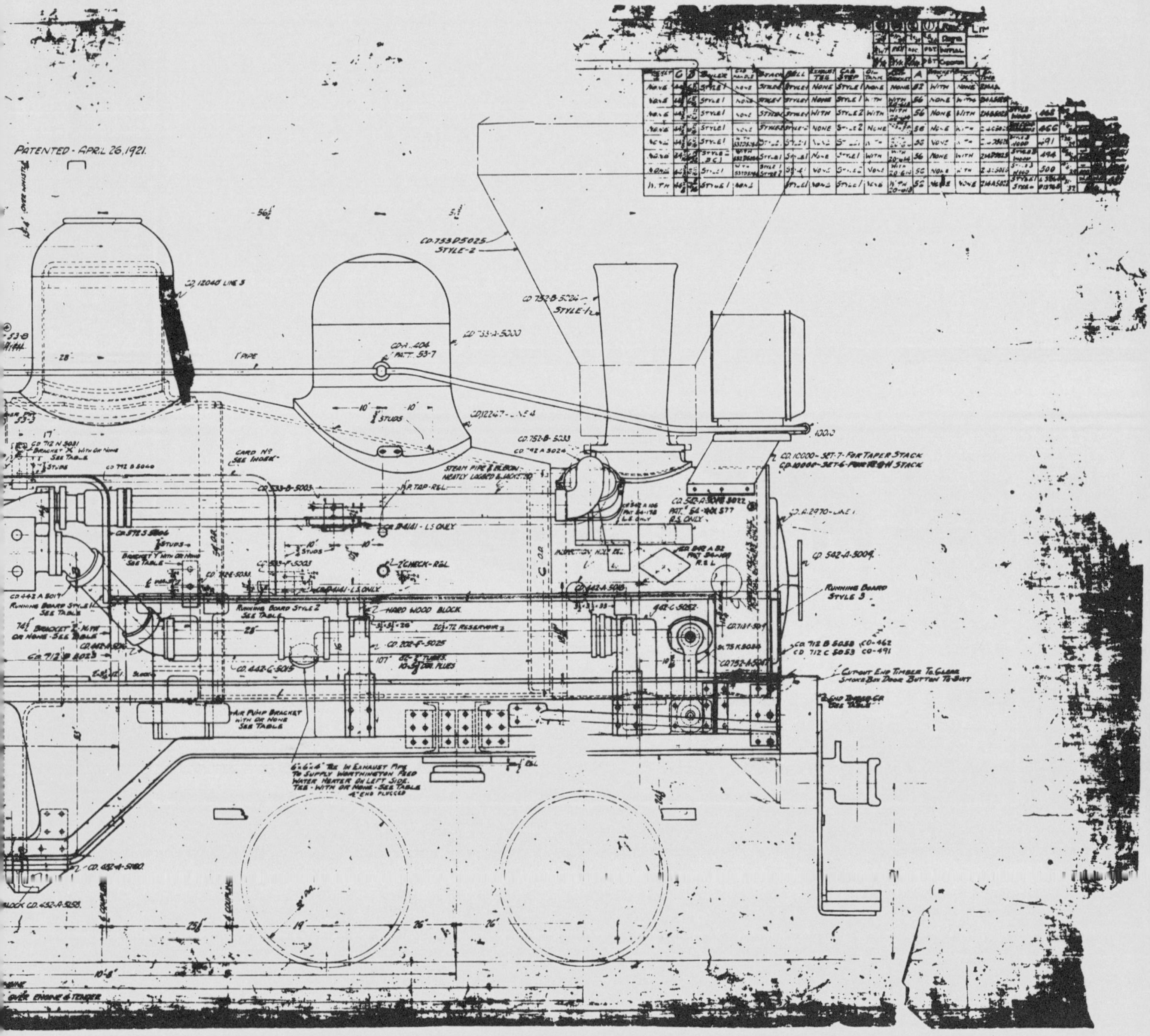

Original Lima drawing of Swayne No. 6. (Authors' collection)

THE SWAYNE LUMBER COMPANY
Narrow Gauge Logging in the Merrimac Forest

LIBRARY OF CONGRESS NUMBER:
91-068501

ISBN: 0-915713-24-1

MAPS:
Bob Mariano
California State Archives
David W. Braun
Ernie Von Ibsch collection

PLANS:
Al Armitage
John E. Robinson
David W. Braun
Herman H. Darr
Anthony G. Wren

ILLUSTRATIONS:
John Coker
Jim Finnell

COVER ART/BOOK DESIGN:
Mike Pearsall

TYPOGRAPHY:
A2Z Graphics
Sedro-Woolley, Washington

LITHOGRAPHED IN HONG KONG

Swayne narrow gauge right-of-way, south of Mooretown Road near Oroville, July 2, 1939. (Bert Ward)

Table Of Contents

Lloyd Swayne was very interested in technology and delighted in photographing both the old and new aspects of his lumber mill and operations. His view of the boiler smokestacks, electrical lines and piping at the northwest corner of the mill complex is almost abstract, yet very mechanical in tone. This is only one of the many photographs from the Swayne family album reproduced here for the first time. (Courtesy the estate of Lloyd Swayne)

The Road to the Merrimac Forest

INTRODUCTION by David W. Braun

A book of history is, by definition, chronological in nature. Add to that the dimension of space and the book begins to assume a shape and form of its own. "SWAYNE LUMBER COMPANY, Narrow Gauge Logging in the Merrimac Forest" has both of these attributes in large degrees. Within these pages are over one hundred years of events, along with a kind of shifting geography, as the reader is taken from the canyon of the Truckee River many miles northward to the Merrimac Forest. Then too, this book travels even further, beyond the end of Swayne Lumber Company, southward to places with names like Pino Grande, South Cable, Tuolumne and Camp 45.

The careful reader will note that beyond the romance of Shay locomotives working in the forests with colorful loggers, this volume is also a journey through the evolution of technology. The events described here make this book a microcosmic panorama of the lumber industry. Here, transportation started with rivers, oxen and chutes, and proceeded to railroads and finally motor trucks. The railroads themselves started with horse-drawn cars, changed to small rod locomotives running on a few miles of track and finished out with sixty-ton Shays running on fifty miles of rail. Skidding technology progressed from oxen and horses, through steam donkey engines to the tractor and arch. The reader will see the reverse as well. With the advent of the cardboard box came the decline of the box shook business, once a mainstay of many lumber enterprises.

Along with the technology, geography and events, we have tried to balance out the book by keeping the human factor in the foreground, by talking about not only the owners, but the employees as well. The reader will be in on some high-stakes boardroom drama and decision-making, and will also get to sit in the locomotive cabs and visit the camps and hear the stories of the men. As will be seen, the owners had to make decisions that would affect the future of their enterprises, and some of these decisions had very long-term ramifications. The crews themselves worked the jobs much like men everywhere — with humor, practical jokes and directness. Some stayed, some moved on, and some were killed.

As business enterprises, the several companies discussed in this book saw the entire gamut of the lumber industry problems; overproduction and low prices, mills burned to the ground, foreclosures, exhaustion of timberlands, loss of vessels at sea, train wrecks, forest fires and the Depression. That Swayne Lumber Company was able to overcome some of these problems was due in part to the resources of the Swayne family and the commitment they had originally made in 1917. That Truckee Lumber Company failed in their attempt to change their base of operations was due in turn to poor management decisions and economic times in general. It may surprise the reader to learn that there were two silent partners involved in all of this, from start-up in 1909 to the very end in 1940 . . . the Southern Pacific and the West Side Lumber Company. The West Side may have been the ultimate winner in this high-stakes game, as they came out of it with 12,000 acres of timberlands, two Shay locomotives and 99 skeleton log cars which permitted them to reorganize their operations and last for another twenty years.

The first eastbound MTy of the day left the Oroville yard at 4:20 a.m. each morning. In the summer, even at that time of day, it was hot as the former dredger tailings on which the entire site was built gave off the heat absorbed from yesterday's sun. The rear brakeman would walk to the Western Pacific bridge where it was a little cooler and wait for the train to come by. He would be "Rollin' 'em by," checking for dragging equipment and locked wheels. The long grade to Summit Siding commenced at the Western Pacific bridge and the Shay locomotive would come by with its throttle wide open.

As the empty skeleton log cars clank by, our man looks 'em over, noting from the glint of the silver paint on the wheels that all are rolling. As the caboose nears, he swings aboard and our train is underway on the main line. Why don't we scramble up after him . . . for our journey to the Merrimac Forest.

DAVID W. BRAUN
Caldor, California
April 15, 1992

Acknowledgements

Over eight years ago I was jeeping in the woods back of Oroville, California and ran across an old narrow gauge railroad grade and then, a trestle. Along with these "finds" and many more to follow, I became interested in discovering everything I could about the Swayne Lumber Company and its narrow gauge railroad. The book you now hold in your hands represents the culmination of eight years of a most interesting and enjoyable endeavor.

Dave Braun, my co-author, has added a quality of depth to this book that would have been impossible without his participation. Dave's talent for ferreting out obscure or difficult information, and for tying up loose ends, is reflected consistently throughout these pages. His rosters, drawings and maps all contribute a great deal to this work, and I look forward to working with him on similar projects.

There are some very special folks who have contributed photographs, information, drawings, help, tips, advice and encouragement. To these people, both Dave and I are greatly indebted: Alan B. Armitage, Reedsport, Oregon; Jim Boynton, Quincy, California; Rich Dunn, Castro Valley, California; Herman Darr, Grass Valley, California; Mallory Hope Ferrell, Peachtree City, Georgia; John E. Lewis, Forest Hill, Maryland; Maggie Nunes, Sacramento, California; U.S. Forest Service archaeologist, Kevin McCormick; Ed Thomas, Oroville, California; John E. Robinson, Denver, Colorado; Russ Simpson, Placerville, California, and David Spohr of Garden Valley, California. In particular, without Mal Ferrell's advice, John Robinson's continual encouragement, Russ Simpson's exceptionally high research standards, and Dave Spohr's thorough knowledge of Truckee Basin lumbering, this book would surely represent a lesser effort.

Several former Swayne employees provided irreplaceable first-hand information that colors this work throughout. To Harry Dennison, a Swayne brakeman and later a Western Pacific engineman; E. E. "Butch" Slaughter, a retired WP conductor and the fireman on Swayne Shay No. 6 from 1928 to 1930; and Walter Sturgeon, who worked as the bullcook at Gold Lake Camp at the age of 19, we offer our gratitude for sharing their memories, good times and bad, with all of us.

Photographs and other help came from many sources and we are indebted to the following people for sharing with us: Kramer A. Adams, San Francisco, California; the Nick Bailey estate; Charles Bates, Allen County Museum; Bob Brown, Los Altos, California; Hart Corbett, Novato, California; Guy L. Dunscomb, Modesto, California; David Garcia, Downey, California; Larry Harrison, Sacramento, California; Ernie Von Ibsch, Oakland, California; George R. Kadelak, Crown Point, Indiana; Don Marenzi, Newark, California; Tom McCaulay, Reno, Nevada; Dave Morgan, Garden Valley, California; Jeffrey Moreau, Fair Oaks, California; the late P.E. Percy; A. C. Phelps, Auburn, California; Steve Polkinghorn, Camino, California; Bill Roy, Eugene, Oregon; Joel Segal, El Dorado Hills, California; John Snyder, Sacramento, California; Dave Squire, Woodland, California; Kent Stephens, Chico, California; Ken Sullivan, Sacramento, California; Dana Supernowicz, Placerville, California; Glenn Tyra, Tuolumne, California; Harold K. Vollrath, Kansas City, Missouri; Bert Ward, Eugene, Oregon; Will Whittaker, Mill Valley, California; Ted Wurm, Oakland, California; Kyle Wyatt, Davis, California.

Robert Swayne Lyons of Santa Barbara, California very kindly made the personal photograph album of his grandfather, the late Lloyd Swayne, available. The photos in this album made the photographic record after 1934 much more complete than it otherwise might have been. Mr. Lyons' interest in our project is very much appreciated.

To Ellen Schwartz at the California Railroad Museum; Joe Samora, California State Archives; Linda A. Mainville, PhD, Nevada Historical Society; and the staff at the Butte County Historical Society; the County Recorder Offices of Butte, Placer and Nevada; Denver Public Library, Western History Department; Sacramento History Center, Archive Section; and the Truckee/Donner Historical Society, our sincere thanks.

Finally, to Eric Bracher who got me started in writing this book . . . thanks! A special thank you is also in order for Eric for commissioning and donating the title page paintings by Jim Finnell and John Coker. Eric began this collection of Swayne photographs, spent time with me exploring the Merrimac Forest, developed kits based on Swayne prototypes, and kept checking on and encouraging my efforts. It was Eric's vision that conceived this book and his interest that helped keep it on track.

—Paul Beckstrom

The Swayne millyard at close of operations in March 1940. The 0-6-0 "A" is at far left, then Heisler No. 5, Shay No. 4, the rotary plow and the other 0-6-0 "B." Log dump is at right center with the main line to the woods at near left. (Ted Wurm)

Dedication

To my wife Lana, for having endured hours upon hours of field trips, typing, telephoning, photocopying, darkroom work and all the other bits and pieces that went into this book.

Also, to E. E. "Butch" Slaughter and Harry Dennison, who worked for Swayne and whose many colorful human-interest stories helped make times past come alive again.

—Paul Beckstrom

To William J. Braun Jr., who encouraged a son's interest in narrow gauge railroads.

—David W. Braun

Early Truckee Basin lumbering scene, believed to be the Truckee Lumber Company. (Nevada Historical Society)

CHAPTER ONE

The Truckee Lumber Company and the Early Years

The Truckee Lumber Company was formed in 1867 by E. J. Brickell and George Geisendorfer. The partners built a water-powered sawmill at Colburn's Station, which became Truckee in August of that same year.[1] The mill's location was predicated in part by the Central Pacific Railroad, then building across the summit of the Sierra Nevada into the western reaches of the Truckee River basin.

The mill was built to satisfy a contract to cut railroad ties and bridge timbers for the Central Pacific (10,000 ties and two million feet of bridge timbers). The total output for 1867 was eight million feet. During the next three years, $150,000 was spent to improve the mill facility. By 1872 the four saws in the mill were cutting at a rate of 50,000 feet every twelve hours.

In 1873 W. H. Kruger purchased a half-interest in the Truckee Lumber Company, renewing a partnership with E. J. Brickell dating back to 1855 in Dutch Flat, California, where the two ran a steam-powered sawmill operation. Brickell was considered to be the lumber and logging expert while Kruger's contribution was in the business end, dealing with contracts and banking houses.

Brickell and Kruger purchased four hundred acres of timberland, which included the western end of Truckee, for a new sawmill site and housing. When the area was subdivided by the company, many of Brickell's family built their homes on the site and the district is still referred to as "Brickelltown."

Brickelltown was built at the bottom of a steep hill dotted with a number of small springs. Most of the homes faced south and looked across the Central Pacific tracks toward the Truckee Lumber Company mill buildings. W. H. Kruger's home remains to this day, bearing the name "White House," after C. B. White, a prominent merchant who purchased the home after the lumber company left town.

Early newspapers advertised Truckee Lumber as "Mfgrs and dealers in lumber of every description. Mining and Bridge timbers, common and finished lumber, matched and dressed flooring and ceiling, bevel edge siding, rustic siding, fancy pointed pickets. Doors, blinds, sash mouldings, fancy front doors. Lath and shingles, transoms, window and door frames and casings made to order. Brackts [sic] and all kind [sic] of scroll sawing done to order."[2]

Truckee Lumber continued to cut large quantities of railroad ties. Customers included the Eureka and Palisade, Utah and Northern, and the Denver and Rio Grande Railway. In 1883 the Truckee mill employed fifty men and had a capacity of 75,000 feet in twelve hours.

Richardson Brothers pole railroad in Martis Valley. The Richardson's mill was at the junction of Martis Creek and the Truckee River. Circa 1872. (California State Railroad Museum)

Other Truckee Lumber Company facilities included a water mill at Alta, a three-story box factory, a fireproof dry house and warehouse in Truckee, and a warehouse, box factory, wharf and general offices in San Francisco, all located within a block of Central Pacific Railroad tracks.[3]

A retail sales outlet was opened in 1885 in Ogden, Utah, called the Sierra Nevada Lumber Association. Logging and milling were done by the Truckee Lumber Company under the "Sierra Nevada" tradename.

During this same time, Truckee Lumber expanded its ocean-going operations beyond simply shipping and receiving at their San Francisco wharf. They established a flag vessel, *The Truckee,* that sailed between a mill at Hobsonville, Oregon and the Bay. The *Truckee* was something of an historic vessel in her own right. Built in 1869 as a barge for the (original) Western Pacific, she was refitted in 1875 as a two-masted schooner under the ownership of her Master, Roger Baker. Baker concluded an agreement with Truckee Lumber to carry their goods under their nominal flag. The agreement was terminated in 1890 when the vessel was moved to another registry.[4]

Over the last thirty years of the Nineteenth Century, Truckee Lumber operated an extensive logging endeavor with mills at Donner Lake, Truckee, Dog Valley and Billy Macks Flat (two miles northwest of Donner Lake). The Truckee mill was the largest, actually being two complete mills built together. Each mill was powered by a low-head water turbine wheel of one hundred horsepower each. The box factory, office and Kruger's home at Truckee had incandescent lighting installed as early as 1890.[5]

From 1883 to 1909 Truckee Lumber operated a number of small narrow gauge railroads incidental to cutting operations. Early newspaper accounts indicate mileages of one and one-half to four and one-half miles in operation at any one time. These lines apparently lasted in a given location only as long as the timber-cutting was taking place, and were then pulled up and moved to the site of new activity.

Original power was furnished by oxen and horses, with several small steam locomotives providing power from 1885 to the end of operations in 1909. The only locomotive ordered new by Truckee Lumber was 0-4-0T Porter c/n 699,[6] built in July 1885.

Truckee's early logging methods were a unique mixture of animal- and steam-powered railways, log chutes and water transportation. The initial cutting areas were located along both sides of the Truckee River between Lake Tahoe and Truckee. Logs were cut on the high plateaus of the Truckee River canyon and skidded by horses or oxen to chutes made from logs. The cut logs were sent down the chutes at high rates of speed, aided in part by grease, to the Truckee River.

The logs were assembled in the river into booms and floated downstream to Truckee, the Truckee River being the outlet for Lake Tahoe. A dam on the Lake Tahoe outlet of the Truckee River would be opened

Crystal Peak Mills office at Verdi, Nevada in 1895. Crystal Peak was a Truckee Lumber Company subsidiary. (Nevada Historical Society) Below, a flat-borromed flume in the Martis Valley belonging to early Truckee area lumberman, George Schaeffer. (Russ Simpson collection)

LTRy&TCo. train dumping logs into the Truckee millpond. SP main line appears in the background. Train is on the LTRy main line. (Russ Simpson collection) Below: An early photo of Clinton Railroad. (Pacific Lumber & Wood) No. 1 Locomotive was Porter No. 147 or 148 built for the Sutro Tunnel Company in Nevada. (Mallory Hope Ferrell)

temporarily to provide adequate water flow for each log run.[7]

Log chutes were extended into the woods along each side of the Truckee River, as far back as any reasonable gradient could be obtained. Average grades ranged between 25 and 45 degrees. The chutes were built of two parallel logs, averaging from 12 to 14 inches in diamater, laid six inches apart and slightly imbedded in the ground. The small end of each chute log was notched into the large end of the log next to it. Chute logs were generally thirty to forty feet long. Twice a day, a boy with a pail filled with warm tallow and a swab would grease the inside surfaces of each chute to reduce friction generated by the logs.

The main reasons for the use of chutes were the proximity of the timber to the river, the slope of the terrain and the inexpensive cost. A narrow gauge railroad cost $4000 per mile, while a log chute could be built for $300 a mile. Four men and four oxen could lay a quarter mile of chute per day. Once the terrain flattened out, the horse, oxen or steam-powered railway was utilized.

The tremendous quantities of lumber cut by the Truckee-area mills peaked out and began declining in 1877-78. The economy of the Truckee River basin lumber industry had developed to satisfy the following demands:

1. The construction of forty miles of Central Pacific Railroad snowsheds, which initially required three hundred million feet of lumber and an annual quantity of twenty million feet for maintenance and replacement.

2. The Nevada silver mining industry, which required seventy million feet annually for tunnel bracing and surface facilities.

3. The construction of new settlements in Nevada and western Utah, which utilized lumber shipped from the Truckee area, since quality lumber at reasonable prices could not be obtained elsewhere.[8]

In 1877 the annual silver output of the Comstock attained an all-time high but decreased dramatically throughout the next four years. Truckee Lumber Company production figures necessarily followed suit.

From 1880 Truckee Lumber began diverting more and more lumber into box-making, converting three million feet into boxes in 1880, seven million feet in 1881 and nine million feet in 1882. A second box factory was built in San Francisco to manufacture boxes which were shipped to points all along the West Coast, including Mexico, Central America and Alaska.

The year 1883 marked a good deal of activity, with Truckee Lumber in control of twelve thousand acres of timber and operating four camps between Tahoe and Verdi. The firm constructed a 1¼-mile narrow gauge line up Washington Creek in Placer County and a 1½-mile line on the east side of the Truckee River, three miles upstream from Truckee. Also in 1883, Dog Valley, near Verdi, was logged by the Truckee Lumber subsidiary, Crystal Peak Lumber Company. Crystal Peak was purchased by Truckee Lumber from Katz and Henry and operated until 1889. In that same year, 2½ miles of narrow gauge were also constructed up Bear Creek, a tributary of the Truckee River, 2½ miles north of Lake Tahoe.

From 1884 until the turn of the century, timber cutting fluctuated but never exceeded ten million feet. Output from 1890 to 1900 was approximately the same as the previous decade.

In 1894 Truckee Lumber completed one of the largest timber trades on record on the West Coast. Five thousand acres of land owned by Truckee Lumber, lying between Truckee and Sierra Valley, were exchanged for five thousand acres owned by the Sierra Nevada Wood and Lumber Company.[9] The new timberlands were located between Truckee and Lake Tahoe.

During the last years of company activities in the Truckee Basin, from 1900 to 1909, operations focused on company lands accessible to the narrow gauge Lake Tahoe Railway and Transportation Company line, which ran along the Truckee River between Lake Tahoe and Tahoe City.

The proposal by the Bliss family to construct the "Lake Tahoe Railway and Transportation Company" railroad between Tahoe City and Truckee instantly attracted the interest of the Truckee Lumber Company. Here was an opportunity to eliminate the expensive river drives and wagon hauls of logs in the Truckee River canyon, and substitute them with a more dependable rail carrier. Here also was the opportunity to further exploit Truckee Lumber land holdings by virtue of the fact that almost the entire right-of-way the proposed railroad would occupy was owned by Truckee Lumber Company.

The historical record is not clear as to who approached whom, but after some discussion between the two parties, a covenant was drawn up on December 13, 1898, whereby the Truckee Lumber Company gave without charge a fifty-foot-wide right-of-way over their canyon properties and agreed to furnish at cost all railroad ties and bridge timbers. The Lake Tahoe Railway in turn agreed to haul logs at a charged rate just over actual cost. This agreement permitted the railroad to start construction in April 1899 and the job was finished in May 1900.

Many years later, during a hearing on the valuation of the LTRy&TCo. before the State Railroad Commission, Walter T. Bliss, in sworn testimony, claimed it was a regular business arrangement:

> "Well, the right-of-way was a part of the consideration; if we had not had the hauling of the logs, we would have not built the railroad."

He was rebutted by a Commission witness, R. A. Thompson, who countered:

Pacific Lumber & Wood No. 2 poses with a log train near Clinton, California (John Robinson collection) and with her crew at Truckee in the summer of 1897. Left to right: Carl Finnegan, fireman; John Finnegan, engineer, and master mechanic Charles Harold. Engine became Truckee Lumber No. 2 and later Swayne "B." (Ted Wurm collection)

Pacific Lumber & Wood train of logs and cordwood crossing the Truckee River on the Donner and Tahoe RR bridge in the late 1890's. Cars are lettered for the PL&WCo. This view looks west with the Central Pacific main line on the upper right. The mill and log dump were off to the right. (Al Graves collection)

PL&W No. 3, the "Dewey," with a log train at the Truckee Lumber millpond sometime between 1893 and 1895. The Central Pacific Railroad main line and spur are in the background. Handwritten location at the top of photo is incorrect. (Al Graves collection)

". . . the right-of-way was in fact donated and consideration as passing between the companies was mutual to the extent they got it from — the railroad company got its right-of-way and the lumber company got a market for its lumber."

During this same hearing, the Commission witnesses were able to show that the railway had reported the land as costing a certain sum, when in fact it had been given to them by an unpublicized covenant; and, moreover, they recorded in their books a far higher freight rate than in fact they were receiving from Truckee Lumber. When confronted with these facts, the Bliss attorney attempted to explain it all away as an oversight:

"Perhaps I should have stated in this matter all the right-of-way the company gave, they entered upon its books what the cost of that right-of-way was in the matter of its acquisition from the Truckee Lumber Company; in other words the special concession in the matter of freight rates, possibly through an oversight in bookkeeping, was not entered properly upon the records. We do not know at this time what that would have amounted to in dollars and cents."[10]

By June 1900 the Lake Tahoe Railway was making two round trips a day from Truckee Lumber's camp at Tahoe City to the mill at Truckee, a distance of fourteen miles. The trains were hauling approximately 75 logs each trip.[11] The hauling of logs by the LTRy&TCo. was done under contract for the Truckee Lumber Company, using Lake Tahoe Railway crews and flat cars. In 1903 the freight rate charged to Truckee Lumber was 23 cents per one thousand board feet measure,[12] although based on the above, the actual rate was somewhat lower.

Despite the murky financial aspects of the business deal, the Lake Tahoe line gave good service to the lumber company, and ultimately in March 1909, for the sum of ten dollars in gold coin, the fifty-foot-wide right-of-way was conveyed to the railroad in perpetuity. For the many miles this right-of-way occupied, the total land area amounted to only 82.6 acres (not including the terminal areas at Tahoe City and Truckee), as compared to the total holdings of the Truckee Lumber Company in the area — thirteen thousand acres.[13]

In July of 1901 Truckee Lumber purchased Pacific Lumber & Wood Company's "Donner & Tahoe Railroad" for the sum total of ten dollars in gold coin. Included in the purchase were two 0-6-0 tank locomotives, 22 flat cars (ten equipped with cordwood racks), three water tanks, a bridge crossing the Truckee River, two car sheds, a freighthouse, an enginehouse and all of the D&TRR right-of-way.[14]

The eleven-mile railroad had been constructed in 1893 by Pacific Lumber and Wood in a southeasterly direction from the Truckee Lumber Company millpond in order to tap PL&WCo.'s timber holdings in the Martis Peak area. The Donner and Tahoe was originally proposed as a common carrier, with plans to extend all the way to Lake Tahoe, carrying freight and tourists, as well as cut logs. These ambitious goals were never realized. During construction, rail and other materials were hauled over the hill by wagon teams from Pacific Lumber & Wood's "Clinton Railroad" at Clinton, ten miles east of Truckee, this latter line being torn up to furnish materials for the D&TRR.[15]

In 1893 PL&WCo. contracted with Truckee Lumber Company to cut one hundred million feet of saw logs to be delivered to the Truckee millpond by the Donner and Tahoe Railroad.[16] By 1901 all available timber access by the D&TRR had been exhausted and the line was dismantled following the Truckee Lumber purchase.

The two Pacific Lumber & Wood locomotives remained on the Truckee Lumber property until close of operations in 1909. The locomotives were used both as mill dry yard switchers and as woods engines on the Ward and Squaw Creek operations. The 22 PL&WCo. flatcars were apparently utilized in the Truckee Lumber dry yards, across the Truckee River from the mill, and many may have eventually made their way to the Butte & Plumas Ry. at Oroville ten years later.

The years 1902 and 1903 marked two major fire losses for the Truckee Lumber Company. In May 1902 the box factory burned to the ground. The factory, a $20,000 loss, was rebuilt over the next few months. In October 1903 the sawmill at Truckee burned, along with fifteen thousand feet of drying lumber. Truckee purchased the Yerington & Bliss mill machinery at Glenbrook, Lake Tahoe and moved the equipment to a new mill structure at Truckee.

It is interesting to note that Truckee Lumber, back in 1883, had organized a plant fire department with a hand-drawn fire engine and hose carriage which was housed in a fire hall located on the mill grounds. These pieces of fire apparatus, along with the mill worker fire fighters assigned to them, were regularly dispatched to fire calls in the town, resulting in the highest praises from the local newspaper, *The Truckee Republican*.

The final years marked the construction of several small rail lines. In 1903 a narrow gauge branch was built from the Lake Tahoe Railway at Tahoe City, two miles south, then west to the head of Ward Creek. In 1904 a new sawmill was constructed at Billy Mack Flat, two miles west of Donner Lake where eight million board feet of timber remained under Truckee ownership. After the Ward Creek timber was cut, another branch rail line was built up Squaw Creek in the Squaw Valley, now a famous ski resort and the site of the 1960 Winter Olympics. A contract was signed with the Scott family, owner of the Squaw Creek property, stipulating that no construction or operations be done in the meadow below the sage line, to hold damage to a minimum.[17] A camp

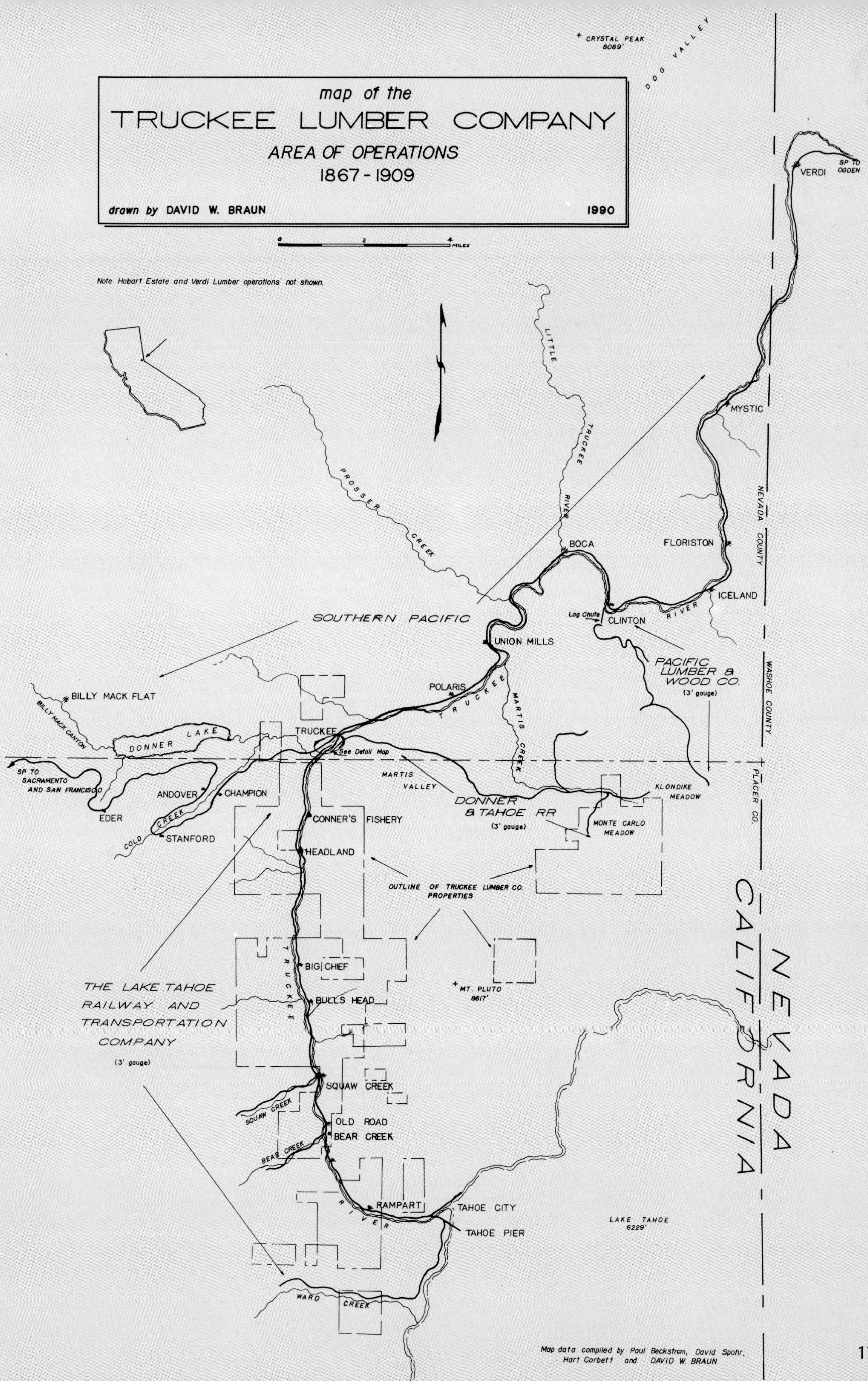
map of the
TRUCKEE LUMBER COMPANY
AREA OF OPERATIONS
1867 - 1909
drawn by DAVID W. BRAUN
1990
0
2
4
MILES
Note: Hobart Estate and Verdi Lumber operations not shown.
CRYSTAL PEAK
8089'
DOG VALLEY
VERDI
SP TO OGDEN
MYSTIC
LITTLE TRUCKEE RIVER
PROSSER CREEK
BOCA
FLORISTON
ICELAND
Log Chute
CLINTON
RIVER
NEVADA COUNTY
WASHOE COUNTY
PLACER CO.
SOUTHERN PACIFIC
UNION MILLS
PACIFIC LUMBER & WOOD CO.
(3' gauge)
POLARIS
TRUCKEE
MARTIS CREEK
BILLY MACK FLAT
BILLY MACK CANYON
DONNER LAKE
TRUCKEE
See Detail Map
SP TO SACRAMENTO AND SAN FRANCISCO
MARTIS VALLEY
ANDOVER
CHAMPION
EDER
COLD CREEK
STANFORD
DONNER & TAHOE RR
(3' gauge)
KLONDIKE MEADOW
MONTE CARLO MEADOW
CONNER'S FISHERY
HEADLAND
OUTLINE OF TRUCKEE LUMBER CO. PROPERTIES
NEVADA
CALIFORNIA
BIG CHIEF
THE LAKE TAHOE RAILWAY AND TRANSPORTATION COMPANY
(3' gauge)
MT. PLUTO
8617'
BULLS HEAD
SQUAW CREEK
SQUAW CREEK
OLD ROAD
BEAR CREEK
BEAR CREEK
RAMPART
TAHOE CITY
TAHOE PIER
RIVER
LAKE TAHOE
6229'
WARD CREEK
Map data compiled by Paul Beckstrom, David Spohr, Hart Corbett and DAVID W. BRAUN

Several views of Truckee, California taken about the turn of the century. At right is the Truckee Lumber Company office with the Central Pacific's main line in the foreground. Below: The west end of Truckee looking west. Large white building in center left is the Truckee Lumber box factory and the dark building at right is the TLCo. warehouse. Below right: Truckee in 1908 with the lumber tramway across the Truckee River in the foreground. (D. J. McDermott collection, Nevada Historical Society and John Robinson collection)

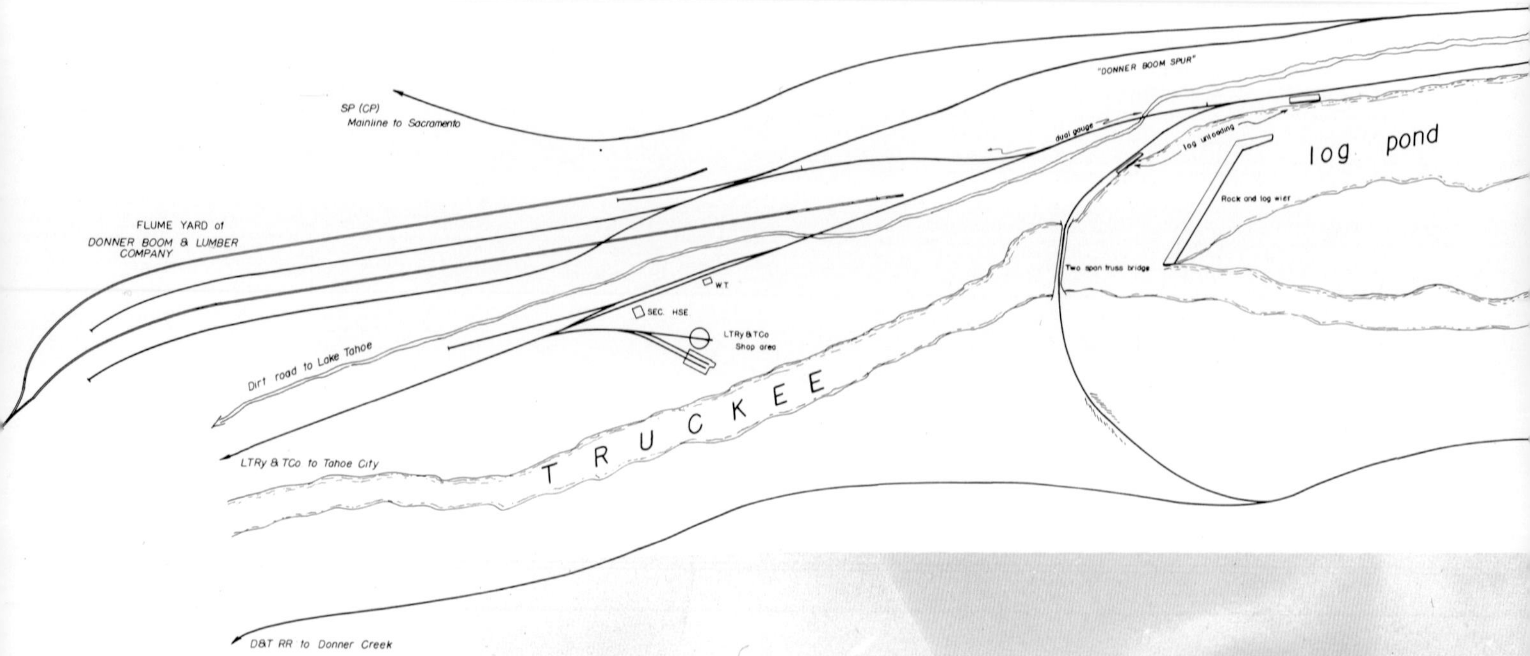

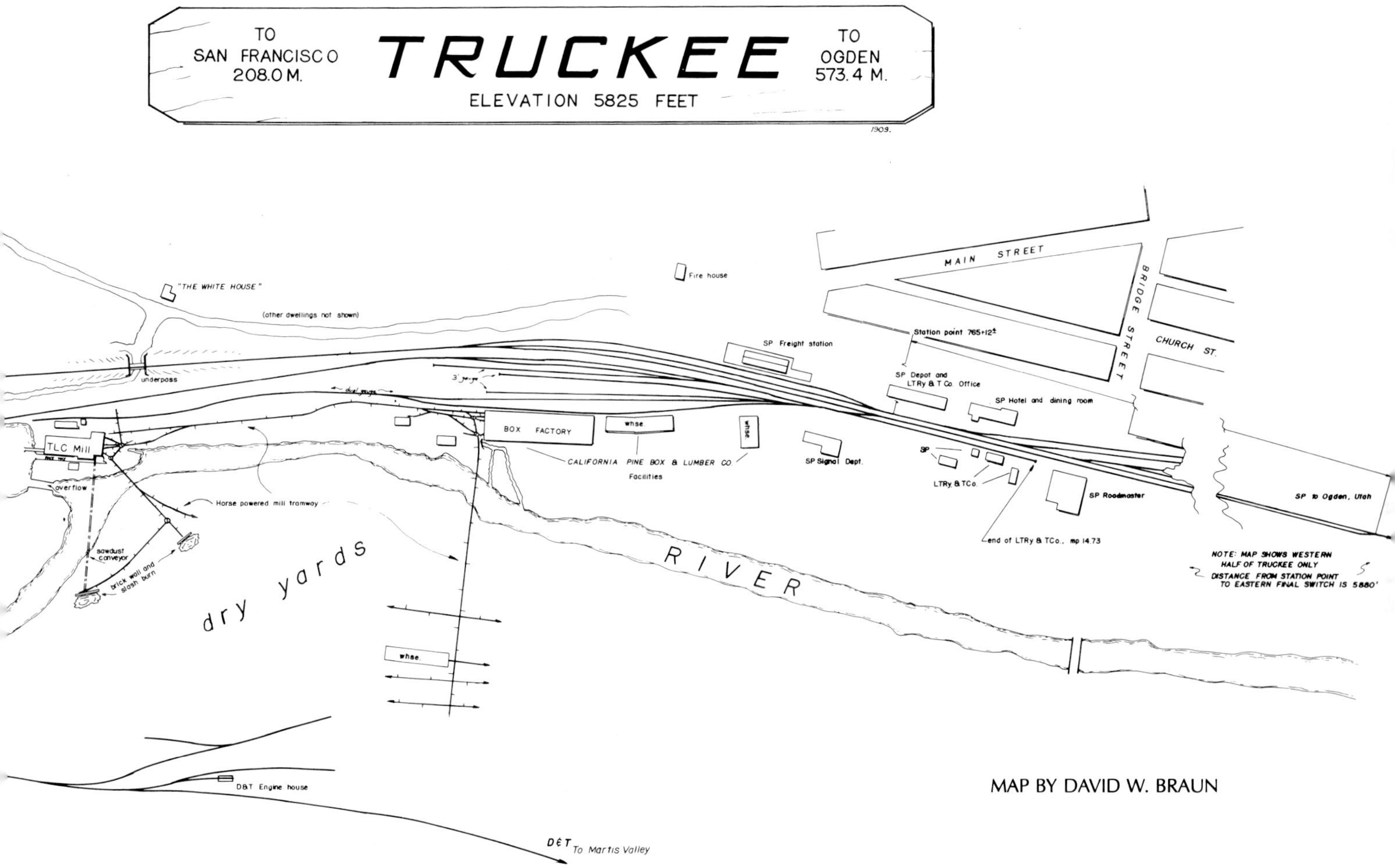

MAP BY DAVID W. BRAUN

The Truckee Lumber Company mill, looking west. Note the PL&W cordwood cars on the dual gauge. Donner & Tahoe RR grade is visible on the hillside across the Truckee River. Horse-powered tramway crosses river to the dry yards. Below: The mill, looking east toward the Nevada State Line, around 1906. (Both photos Russ Simpson collection)

was established in the valley in 1907, and the logs again were hauled under contract by the Lake Tahoe Railway. The Ward Creek Logging Railroad was sold to the Sierra Realty Company in December 1909, including rails, ties and most buildings.[18]

The final day of business for Truckee Lumber Company at Truckee was November 6, 1909 and it was the subject that day of a warm farewell from *The Truckee Republican:*

> "The Truckee Lumber Company was organized in 1866, and since that time has operated continuously, employing thousands of men and animals, and distributing hundreds of thousands of dollars in wages and for supplies. About two weeks ago, the work of dismantling the lumber camps was begun, and men and teams sent over the hill.
>
> "And now the institution that has supported so many happy families in the past is now silent — a memory."[19]

The process of disposing of the various properties in the general Truckee area had begun earlier that year but proved to be an undertaking of many years duration. The largest piece of real estate was, of course, the Truckee River canyon area. After failing to find a buyer for the entire property, Truckee Lumber hit upon a novel (for the times) idea: subdivide the river front into lots and sell them to individual buyers, for summer homes, cabins and trout-fishing.

To avoid various liabilities, the entire property was sold to a strawman, H. L. Lanfar, in December 1909 for the usual ten dollars in gold coin.[20] In January 1910 real estate men were brought in, including Wilson & Lindsay of Oakland, Wickham Havens of Piedmont and the Bliss-owned Sierra Realty Company. H. L. Lanfar herself was a secretary for the Truckee Lumber attorneys in San Francisco.[21] Landowner Lanfar formed the Tahoe-Truckee Forest Company which ostensibly the real estate men were representing, but it evidently was well known around Truckee that the stockholder names at both Truckee Lumber Company and Tahoe-Truckee Forest Company were one and the same.[22]

The real estate people employed some heavy-handed tactics, including posting of the property against trout fishing, and by May 1910 *The Truckee Republican,* lately the friend of Truckee Lumber, was running diatribes against the company accusing it of:

> ". . . running a bluff to compel people to purchase lots in order to enjoy fishing privileges along the river as it claimed each purchase will thus purchase exclusive right to fish."[23]

Saying the only way the company could keep people out would be through the use of a barbed-wire fence, the *Republican* dared them to do so, then answered its own challenge:

> "This they will not do, thusly the fisherman can fish until dooms day, [sic] as far as being molested by the agents of the Truckee Lumber Company."[24]

Due to these and other factors, the lots did not sell very well. H. L. Lindsay later testified in the 1913 hearing that the actual selling prices were anywhere from one-half to two-thirds of the published price; and Walter T. Bliss, in the same hearing, had to admit that only 59 of 610 lots had sold by that point, and blamed it on, as he put it, "tight money." Whether that meant general economic conditions or more simply that people were not rushing to buy into the Truckee River canyon was not made clear.[25]

To complicate matters even further, the Lanfar contract was written in an attempt by Truckee Lumber to reserve so many rights, including water and riparian, that the resulting document was almost unenforceable as a legal instrument for subsequent sales or transfer of the land. In 1944 the U.S. Forest Service, during negotiations for a timberland swap in the area, found the titles so cloudy that quit-claim deeds finally had to be requested of the West Side Lumber Company (a Truckee Lumber lienholder), the Southern Pacific (successor to the Lake Tahoe Railway) and the Sierra Pacific Power Company of Reno. At that, the USFS admitted there legally was a "dual ownership situation" over which they had little actual redress.[26]

Part of the Lanfar contract reserve clauses centered about the waters of the Truckee River. In this instance, the Truckee Lumber Company was trying to anticipate a future development of the Southern Pacific, that of the electrification of the "Hill," the SP main line over Donner. Periodically, "talk" about the electrification project surfaced and it so happened in the 1909-10 period that there was a reasonable expectation that the Southern Pacific would proceed with this development, along with the double-tracking project they were engaged in at the time. Truckee Lumber thusly held onto its water rights, hoping to either sell or lease them, or perhaps even participate in a water-powered electric plant or generating station.[27] Ultimately, as the electrification project did not come to pass, Lanfar sold the water and riparian rights to the Truckee River General Electric Company, which was a predecessor of the present-day Sierra Pacific Power Company in the same area.

At the time, of course, much of this was in the far future. With the cutting of the last logs and the closing of the mill in November 1909, the relocation of the company to new timber holdings seemed imminent. The move was actually part of a well-planned scenario that had been in the works for some time.

As early as the mid-1880's, William H. Kruger of Truckee was looking for other areas to expand his lumbering operations. With his timber holdings in the Truckee/Tahoe area being depleted, Kruger's attention settled on the vast Merrimac Forests in Butte and Plumas

After the TLCo. posted its Truckee River canyon properties against trout fishing, some of the local fishermen decided to be a little more circumspect. Here, an angler tries his luck from a Lake Tahoe Ry. and Transportation Co. bridge north of Tahoe City, seeming to prefer outrunning a LTRy&TCo. train rather than face an agent of the Truckee Lumber Company. (David W. Braun collection)

counties, northeast of Oroville, California. In 1899 the Merrimac Land and Lumber Company was incorporated in Plumas County with W. H. Kruger as one of the directors.[28] The purpose of this company was to purchase timber holdings in Butte and Plumas counties, especially in the area surrounding the village of Merrimac. The Merrimac district is bounded on the north by the Feather River canyon and by the Middle Fork of the same river to the southeast. From west to east, the area extends from Berry Creek to Granite Basin and Buck's Lake.

The Merrimac district was the site of a much earlier steam tractor operation. George Gable of Oroville operated in this area in the mid-1880's, using an imported English-made six-ton tractor.

This was not the first nor the last time Kruger ventured beyond the Truckee River basin. In March of 1907 the Truckee Lumber Company purchased the F. H. Bardshar mill and narrow gauge tramway located at Cantara, California on the SP's Shasta Route. Cantara, located just north of Dunsmuir, marked the 18th crossing of the Sacramento River and was inside a large loop formed by the climb of the SP main line toward the summit south of Weed. The narrow gauge railroad ran underneath the SP toward the west. Operations included a switchback to gain altitude on the run towards Castle Lake. It is not known if equipment was supplemented by engines or cars from Truckee, or if any of the Cantara equipment came to either Truckee or the Merrimac operations. The property itself was lost to foreclosure in the Spring of 1912, along with the rest of the Truckee Lumber Company operations.[29]

The Kruger family owned the majority of stock in the Merrimac Land and Lumber Company. In 1901 Kruger included his son-in-law, O. C. Haslett, also with the Truckee Lumber Company, in the venture. The plan was to wait until the Truckee/Tahoe holdings were exhausted and then move the mill and railroad equipment to Butte County. Figuring prominently in all of these plans was the coming of the Western Pacific Railroad to the Feather River canyon. The otherwise inaccessible Merrimac Forest could then be logged economically and the logs shipped to a new mill in the Oroville area.

With the completion of the Western Pacific to Big Bend in May 1908, Haslett, now head of the Merrimac Land and Lumber Company due to the death of Kruger, started surveys to examine the lands in the French Creek and Granite Basin areas for the feasibility of building a logging railroad to connect to the WP. Several months later Truckee officials announced plans to construct a standard gauge line from a point on the Western Pacific near French Creek, up the French Creek drainage to Stanwood, and from there into the Merrimac area.[30]

Several historical events came into play during these developmental years. The first was the 1906 San Francisco earthquake. The *American Lumberman* for July 1906 reported the following:

> "Merrimac Land and Lumber Company, which owns over 12,000 acres of the finest timberland in Butte County, has placed on record at Oroville, a mortgage for $180,000 on its holdings in order to raise funds for the development of the properties. The cutting of the timber has been rendered feasible by the construction of the Western Pacific Railroad. This will allow bringing lumber to the San Francisco

This 1902 map depicts the Butte & Plumas Railway's "paper railroad." (California State Library)

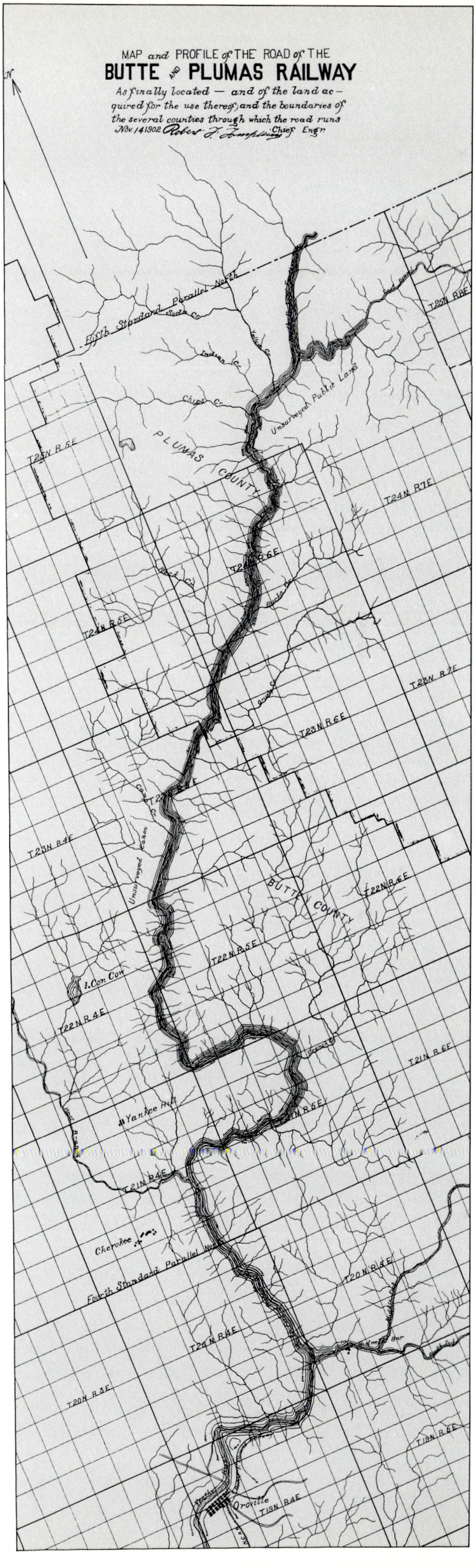

market, where rebuilding the city will require vast quantities for years to come."[31]

The other item of note was the existence of a "paper railroad" in the Feather River canyon during the period 1900 to 1910. This railroad was called the Butte & Plumas Railway.[32] A survey map was filed with the State of California on November 14, 1902, showing a railroad route going from Oroville, up the Feather River canyon and terminating in unsurveyed lands in Plumas County. Apparently the Butte & Plumas Railway was incorporated in 1902 as a front to disguise the Western Pacific's plans to build a line up the North Fork of the Feather River. Truckee Lumber Company bought the name "Butte & Plumas Railway" from the Western Pacific and turned the "paper railroad" into a reality.

By 1909 activity was underway to facilitate the relocation of Truckee Lumber Company to Oroville. In an effort to encourage Truckee to build their mill at Oroville, the Oroville Chamber of Commerce raised $3000 by popular subscription to be applied as a down-payment on seventy acres of land that bordered on the Oroville-Nelson Cutoff Road. Parcels on the seventy acres were purchased from the Western Pacific, El Oro Dredging Company, the Western Ice and Fruit Cooling Company and several individual parties.

Oroville officials believed that it was good business to court the Truckee Lumber Company relocation because the new mill would employ three hundred men, plus another one hundred in the woods, with an anticipated payroll of between fifteen and twenty thousand dollars a month. Besides the primary mill, Truckee planned a box factory and a sash-and-door mill. In addition, the Chamber of Commerce elicited a promise from Truckee Lumber that the firm would not maintain a company store, a common management device at the time to ensure a docile workforce and low wages.

In May 1909 Truckee Lumber Company purchased considerable additional timber holdings along Western Pacific right-of-way in both Butte and Plumas counties and from Joshiah Bartlett. Bartlett had purchased these lands at auction to satisfy a judgment against the original owner, H. H. Yard.[33]

The summer and fall of 1909 was marked by feverish activity in the French Creek area, as surveys were made and preliminary grading completed for the Butte & Plumas Railway. On September 1, 1909 the Truckee Lumber Company capitalization was increased from $300,000 to $1 million and the bonded indebtedness to $600,000. Liens were given for their property holdings in Nevada, Butte and Plumas counties. The money was to be used to pay for construction of the Oroville mill and factory and the Butte & Plumas Railway. Union Trust Company of San Francisco was the trustee for the sixteen-year gold bonds used to raise the money.

Following the September 1, 1909 financial agree-

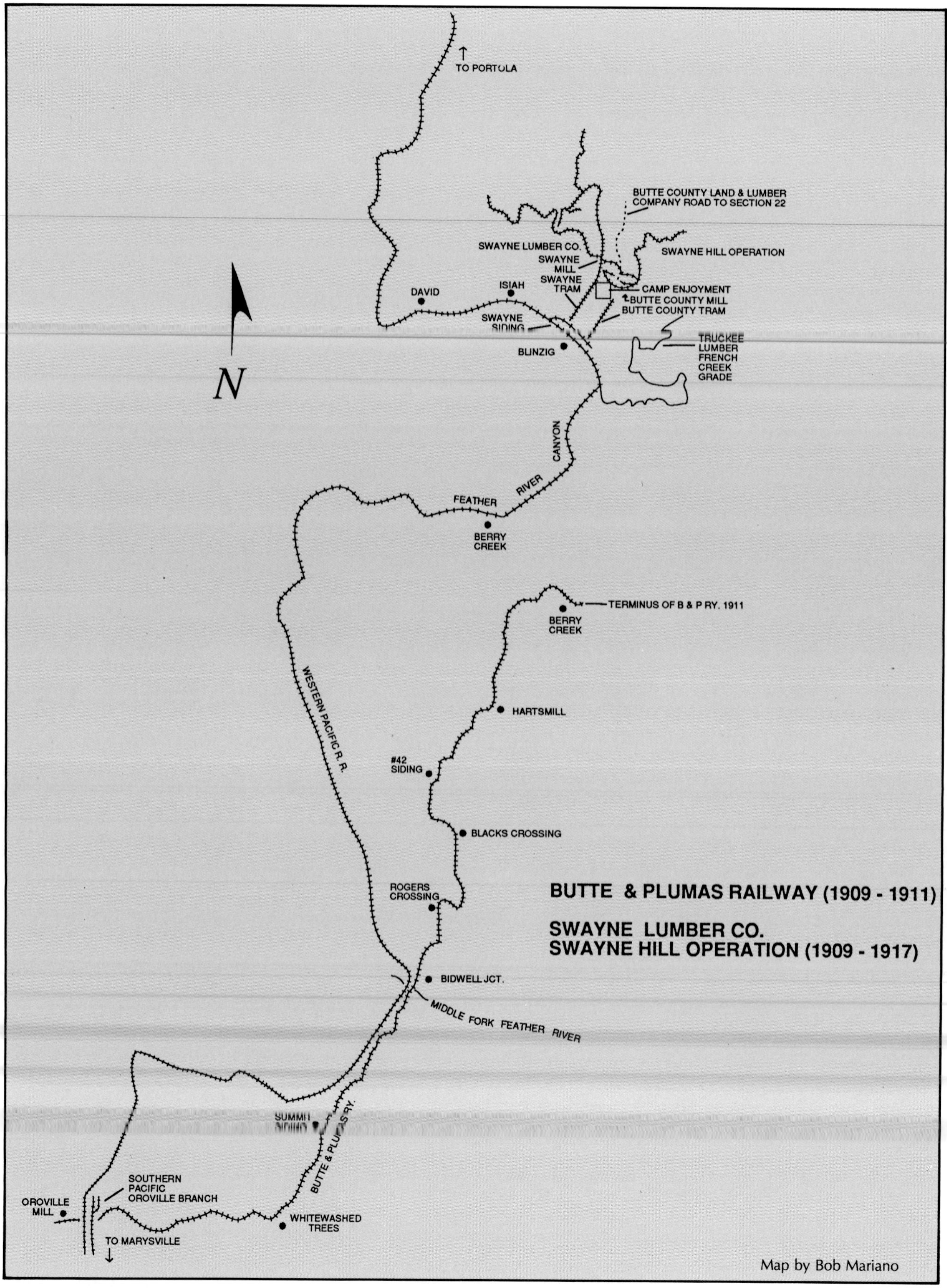
TO PORTOLA
BUTTE COUNTY LAND & LUMBER COMPANY ROAD TO SECTION 22
SWAYNE LUMBER CO.
SWAYNE HILL OPERATION
SWAYNE MILL
SWAYNE TRAM
CAMP ENJOYMENT
BUTTE COUNTY MILL
BUTTE COUNTY TRAM
DAVID
ISIAH
SWAYNE SIDING
BLINZIG
TRUCKEE LUMBER FRENCH CREEK GRADE
N
FEATHER RIVER CANYON
BERRY CREEK
TERMINUS OF B & P RY. 1911
BERRY CREEK
WESTERN PACIFIC R.R.
HARTSMILL
#42 SIDING
BLACKS CROSSING
ROGERS CROSSING
BUTTE & PLUMAS RAILWAY (1909 - 1911)
SWAYNE LUMBER CO.
SWAYNE HILL OPERATION (1909 - 1917)
BIDWELL JCT.
MIDDLE FORK FEATHER RIVER
SUMMIT SIDING
BUTTE & PLUMAS RY.
SOUTHERN PACIFIC OROVILLE BRANCH
OROVILLE MILL
WHITEWASHED TREES
TO MARYSVILLE
Map by Bob Mariano

Glenn Tyra Collection

CAPITAL

SHARES AT A SHARE

San Francisco, Cal., Sept 1st 1909 Number 8

THIS CERTIFIES that Kruger Estate Co 1498 Shares

is the owner of Fourteen Hundred Ninety eight Shares of the Capital Stock of

Truckee Lumber Company

Transferable on the books of the Company by endorsement hereof and surrender of this Certificate

SECRETARY INCORPORATED PRESIDENT

ment, additional construction ensued on the French Creek railroad route. A number of Greek laborers were employed in this operation and the work entailed the used of a steam shovel and considerable blasting. It was during this construction period that two workers were killed. George Spanders, a Greek, and a man named George Dugan from Oakland, were both killed while preparing a blast near the construction camp at French Creek. Newspaper reports stated the men's bodies were blown a distance of two hundred feet.[34]

By late September a contract was awarded to the company of Klippel and McLain to build the standard gauge logging railroad from the mouth of French Creek to Stanwood. This same company had worked for the Utah Construction Company in the building of the Western Pacific Railroad. Work on the railroad continued until March 1910. Three construction camps were set up between the mouth of French Creek and Stanwood. Besides the surveying and grading work, a small sawmill was operated at Stanwood for the purpose of cutting ties and bridge timbers. Fifty men were employed in the milling and logging operations necessary for construction. Contractors estimated that more than twenty thousand board feet of timber would be needed for trestles alone.

Truckee Lumber had leased the majority of the buildings in Stanwood from the Pacific Gas and Electric Company. Stanwood, a small village at the confluence of Jack and French Creeks, was built by the Bay Counties Power Company in 1903 and served as the original construction camp for the ill-fated French Creek Dam project. Truckee survey crews stayed at the boarding house that had been used by power company workers and Stanwood became the Truckee Lumber Company's headquarters for its woods operations.

While all of this activity was taking place in Stanwood and along French Creek, construction was also taking place at the Oroville mill site. The *Pioneer Western Lumberman* offered the following description in January 1910:

> "The Truckee Lumber Co. has prepared plans for its large building on its site at Oroville, Cal., One building will be 200 feet by 70 feet, three stories high, and will be built of steel. The first floor will be devoted to the shafting and motive machinery; the second floor to sawing, and the third floor to saw filing.
>
> "The power building will also be of steel and will house boilers and engine producing 1,050 horsepower. The machine shop is being built of corrugated iron and is nearly completed. An office building of concrete with a tiled roof, mission style, 60 feet by 40 feet, will be commenced soon. After these buildings will have been completed a box factory and planing mill will be built.
>
> "The dam for the millpond and the leveling and scraping of the bottom of the millpond were begun the early part of December. The top of the dam will be utilized as a portion of the Western Pacific spur track.
>
> "The Truckee Lumber Co. will have its own railroad for the transportation of its products.
>
> "It is proposed to run the sawmill all the year. To insure this, the millpond will have a storage capacity of 7,000,000 to 8,000,000 feet. The company expects to begin sawing about July 1, 1910."[35]

The Oroville mill did not begin capacity operation until May 1911, almost a year behind schedule. The reason for this delay was a disagreement between the Western Pacific Railroad and Truckee Lumber Company regarding the hauling of logs over the WPRR from the Butte & Plumas at French Creek to the Truckee mill at Oroville, a distance of thirteen miles. Evidently the agreement between the two companies had been verbal only and, after second thoughts, the WP sought an "out" by demanding more money to haul B&P cars on the main line. The WP cited problems with liability involved with operating log cars on their line.

This unexpected turn of events caused work to be immediately suspended on the French Creek grade. In January 1910, a $400,000 lawsuit was filed by Truckee against Western Pacific to cover costs involved with the construction of three miles of B&P right-of-way up the French Creek grade to Stanwood.[36] For the short term, however, the situation left Truckee Lumber with a nearly completed mill and no way of getting logs to it.

Deciding further negotiation with the Western Pacific was pointless, the Truckee Lumber Company began discussion with the Southern Pacific Railroad. These talks centered around financial assistance for a new railroad to be constructed from mill at Oroville to Hart's Mill. This new railroad would be primarily a captive line to the mill and would not involve reloading or traveling over either existing WP or SP trackage, permitting log cars to travel directly from forest to mill. Early on, it was decided for reasons of economy to build the line to 36-inch narrow gauge.

Apart from being Western Pacific's archrival, Southern Pacific had substantial reasons of its own to cooperate with the Truckee Lumber Company. The Central Pacific — and later the Southern Pacific Land Company — owned a large amount of timberland in the Merrimac Forest. Much of this property was contiguous with that owned by the Truckee Lumber Company.[37] The Southern Pacific also had their Oroville Branch, running 25 miles from Marysville to Oroville with daily mixed service. This trackage was originally built as the California Northern Rail Road, reaching Oroville February 9, 1864. The Southern Pacific Oroville Branch ran parallel to and just east of the Western Pacific main line at the eastern edge of the Truckee mill site. The potential of having a new shipper on the SP branch must have held some importance in the decision to help Truckee Lumber proceed with a new railroad.

The Truckee Lumber/Southern Pacific talks resulted in the Southern Pacific advancing the Butte & Plumas Railway Company, now a common carrier line, $50,000 in June 1910 and an additional $350,000 soon thereafter. Capital stock of the railroad was listed at $500,000 and Oroville was listed as the principal place of business. The B&P Board of Directors included O. C. Haslett, E. S. Dunbar and J. L. Smith, all of Oroville. Of the $30,000 paid into the new company by the board, Haslett contributed $29,800 and the other two men paid $100 each for one share apiece.[38] Despite the local names, rumors abounded that the Southern Pacific was the real owner of the new railroad. These rumors were fueled anew when SP employees were used in several capacities during subsequent construction.[39]

The newly acquired financing allowed Truckee Lumber to commence building a 36-inch narrow gauge line northeast of Oroville along the route following Dry Creek to Saddleback near Whitewashed Trees. It then was to drop down Bidwell Hill into the Feather River canyon, crossing the South Fork of the Feather at the low point near Bidwell Bar, gradually climbing again along Canyon Creek to Berry Creek to the beginning of the Truckee Lumber Company timberlands. As we shall see later, the topography of the Feather River canyon was to set the railroad's operating parameters until the end in 1939, with doubling and tripling of loads on the Bidwell Hill Grade becoming the norm early on.

Realignment of the Butte & Plumas Railway represented a radical departure from the railroad as originally planned. Aside from the obvious change in track gauge, the route was altered to follow drainages far to the south of French Creek. Therefore, rather than a three-mile haul to another railroad, 23 miles of winding, steep right-of-way would have to be traveled to reach any harvestable timber. And, this railroad was to be built and operated by a company whose only previous railway experience involved essentially light tramways with small cars and locomotives.

The Butte & Plumas was completed to Berry Creek by May 1911. Grading was accomplished by horses, Fresno scrapers, a steam shovel and blasting. Again, Greek laborers were employed, as well as a crew of Italians. In all, three hundred men were involved in the construction of the line. Machinery and supplies arrived in boxcars from San Francisco and Stockton. Construction camps were established at Fitz Rogers' place near Bidwell Bar, at Hart's Mill and at an unknown intermediate location. The chief engineer for the B&P was E. S. Dunbar and the superintendent of construction was Stoneman Dorn.

By 1911 operations were in full swing at the Oroville mill and on the railroad. The mill employed one hundred men and was cutting 125,000 feet of lumber daily. Another one hundred were employed in timber-cutting operations near Berry Creek. Finished lumber was shipped from Oroville to yards of the Sacramento Valley Lumber Company, formed in 1910 to serve as Truckee Lumber's chain of retail outlets.

Early on, railroad operations on the narrow gauge Butte & Plumas assumed a reputation for sharp curves and steep grades. On June 18, 1911 a locomotive and

Only photo of a Butte & Plumas train known to exist. Engine is the No. 3 with original wood cab and wood-burning stack. Flatcar No. 12 is lettered "T.L.CO." for Truckee Lumber. Location is Hartsmill. (J. E. Boynton collection)

The original Butte & Plumas bridge across the middle fork of the Feather River around 1909. View looks west toward Land. Western Pacific trestle is in the background. (Ed Thomas collection)

Butte & Plumas No. 4 is shown in this early West Side view, moving a camp setting. This locomotive ran on the B&P a grand total of two months. (J. E. Boynton collection) At right: This map of the TLCo. mill in Oroville (Circa 1911) shows Northern Electric, Western Pacific and Southern Pacific trackage, but does not show the Butte & Plumas Railway which followed the banks of Dry Creek to the east. (Ernie Von Ibsch collection)

four loaded cars plunged off O'Rourke Trestle near Berry Creek. The train fell forty feet to the creek bottom, wrecking the locomotive and cars. The engineer, fireman and head brakeman jumped to safety, suffering only minor bruises and engineer Griffiths breaking his arm.[40]

This potentially lethal accident served to polarize the railroad employees. Claiming the railroad was one of the most dangerous in the country due to its many curves, high trestles and 4½-percent grades, the brakemen and conductors in July of 1911 asked for a reduction of the workday from twelve to ten hours, in return for the conditions prevailing. The request was ignored by the Truckee Lumber Company. The standoff, in turn, resulted in a strike on September 12, 1911 of twenty trainmen, mainly brakemen and conductors, along with a few sympathetic locomotive engineers.

The strike prodded company officials, represented by Everett St. John Dunbar, to meet with the strikers, who were led by H. M. Holdren and F. E. McDurimick, the next day on September 13th. The meeting went nowhere as neither side was willing to compromise. Dunbar then tried airing the issue in the local newspaper, but in an interview printed September 15th in the *Oroville Daily Register,* he did not mention the basic questions about safety, but instead tried to make the issue one of hours worked:

> "Mr. Dunbar in conversation with a representative of the *Register* last evening said that the immediate trouble grew out of the unloading of logs (by trainmen) from cars at the millpond, which custom is followed on all logging roads. He stated that the Butte & Plumas road will be run as a logging road and not as a main line.
>
> "Many of the men in the employ of the Butte and Plumas have formerly been employees of the Western Pacific and the Southern Pacific, he states, and felt that the prevailing conditions on these roads should be effective on the narrow gauge here. They considered it an imposition that they should be required to do the work expected of them here."[41]

The standoff continued and management took the predictable course. An ultimatum was issued: "Return to work or have your jobs filled by new hires." The strikers refused and, as the trainmen's jobs began to be taken by outsiders, violence broke out. A number of altercations occurred and a B&P trainmaster D. G. Lane received a badly lacerated lip and nose from an intoxicated striker. In another confrontation, A. J. Meashan, an official with the B&P, was arrested for carrying a concealed pistol and subsequently leveling it at the strikers.

After a week on strike, most of the trainmen gave in and returned to their old jobs without any of their demands being met by the B&P. At the time, it was said that the strikers had very little clout due to the availability of replacement workers, and the impending onset of winter, which would shut down logging and railroad operations, in any event.

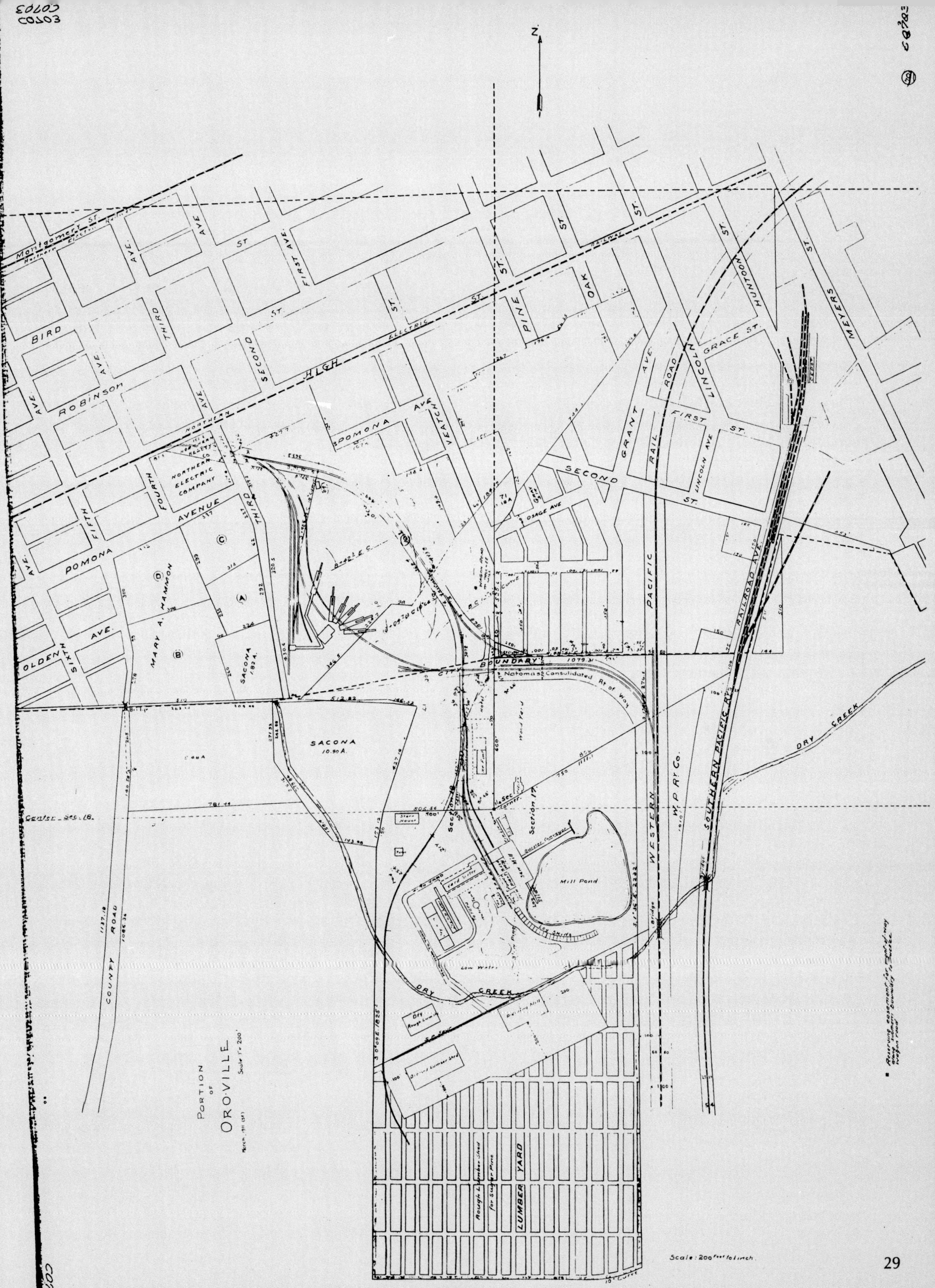

Portion of Oro-Ville
Scale 1" = 200'
March 1911
N
Montgomery St.
Northern Electric Railway
Bird
Robinson
Ave.
Third Ave.
First Ave.
Second St.
High St.
Northern Electric
Pine St.
Oak St.
Pomona Ave.
Veatch
Grant Ave.
Rail Road
Lincoln Ave.
Grace St.
Huntoon St.
Meyers St.
First St.
Second St.
Osage Ave.
Fourth Ave.
Fifth Ave.
Sixth Ave.
Pomona Ave.
Avenue
Northern Electric Company
Mary A. Hammon
Sacona 1.82 A.
Sacona 10.90 A.
City Boundary
Natomas Consolidated Rt. of Way
Western Pacific
W. P. Ry. Co.
Southern Pacific Ry.
Railroad Ave.
Dry Creek
Center Sec. 18.
Section 18
Section 17
County Road
Mill Pond
Low Water
Dry Creek
Lumber Yard
Rough Lumber Shed for Sugar Pine
Scale: 200 Feet to 1 inch.

Ironically the mill and railroad shut down at the end of September 1911 and for good, as far as the Truckee Lumber Company was concerned. After all the effort involved in building a new mill and railroad, seeking finances, fighting the Western Pacific Railroad and unruly employees, the economics of 1911 mitigated against success. Market conditions prevailing at that time resulted in large amounts of unsold lumber in West Coast lumber yards, and a subsequent downturn in sawmilling operations. In part, the Truckee Lumber Company contributed to this glut of cut lumber by setting records in sawmilling operations in 1912, including a California state record for cutting pine timber on a single band saw — 103,417 board feet in ten hours of work.[42]

At the time of the shutdown, Truckee officials said operations would resume in the Spring of 1912, but the Southern Pacific quickly moved to protect its investment, and during the winter of 1911-12 obtained the Butte & Plumas on a default judgment. Some railroad equipment was removed from the property during this time but, by and large, the entire operation sat idle from the Fall of 1911 until the arrival of the Swayne interests in the Spring of 1917.

During that same period, both the Truckee and Southern Pacific concerns attempted to salvage their respective investments by offering the properties to potential buyers. Prominent among supposed buyers was the West Side Lumber Company of Tuolumne, California which operated an extensive logging railroad operation under the Hetch Hetchy and Yosemite Valleys name. The *Truckee Republican* reported the sale to West Side Lumber in its August 22, 1912 issue:

> "G. W. Johnson, Vice President for the Westside [sic] Lumber Company at Tuolumne, announced they have taken over the Truckee Lumber Company at Oroville, formerly controlled by the Haslett interests. The building of the Butte and Plumas Railway is said to have crippled the Company. Truckee Lumber Company has been idle for some time. The Westside [sic] Lumber Company officers are W. R. Thorsen, President; G. W. Johnson, Vice President; and J. R. Prince, Sec . . ."[43]

The circumstances surrounding the sale remain a matter of some historical speculation. West Side was said to have been involved with Truckee Lumber as early as 1910 when they invested $1,000,000 in the Truckee timber holdings. The 1912 "sale" shown above is not clear as to whether this was an outright purchase, a stock purchase or a foreclosure. Whatever, there definitely was a relationship between the two companies as evidenced by a letter from the West Side Lumber Company to a stockholder in 1924, referring to the West Side Properties Company as being organized "to take over a contract that the West Side Lumber Company had with the Swayne Lumber Company of Oroville." The amount of the contract was listed as "$750,000 to be paid in installments of $75,000 a year for ten years without interest." A notation was made that, as of April 1924, no payments had been made.[44] It is believed that the reference to Swayne is an error (since they were very much in business in 1924) and the writer of the letter had confused the Swayne and Truckee separate interests. In any event, West Side Lumber Company certainly felt someone in the Oroville area owed them a large sum of money.

If West Side had invested in Truckee prior to September 1911, then it becomes obvious they took a loss on the investment. This, in part, might explain why Butte & Plumas No. 4 — a big, brand-new, sixty-ton, three-truck Shay — was sent to Tuolumne in May 1912. As the more familiar West Side No. 7, West Side employees often referred to the locomotive as "George W. Johnson's Million Dollar Engine." It was perhaps the only tangible result of West Side's investment into the affairs of the Truckee Lumber Company.

West Side Lumber Company did have the foresight in the Spring of 1912 to acquire by lien approximately twelve thousand acres of timberlands in the general Truckee River canyon area, including some of the disputed Lanfar property discussed earlier.[45] That West Side Lumber was able to acquire these lands lends credence to the view that Truckee Lumber and West Side had some sort of business arrangement dating back to at least 1910, and perhaps earlier. The exact nature of the relationship is unknown, but it obviously did place West Side as a primary creditor.

As a long-term investment, the Truckee River property did finally pay off. Starting in 1942, and extending until 1944, West Side and the U.S. Forest Service had a series of discussions about swapping these lands for ones closer to West Side operations in the Stanislaus National Forest. The arrangement was finalized after some of the legal questions relating to Lanfar were cleared, as discussed previously. As noted in Allan Krieg's book on West Side — *Last of the Three Foot Loggers* — this swap was actually part of a much larger overall trade, totalling over 65,000 acres of Northern California properties.[46]

So far, very little has been said of the Swaynes, the principal subject of this book. But, as will be seen in the next chapter, the Swaynes were busy with their own logging enterprise in the same general area as the Truckee Lumber Company operations. As events ran their course, the Swaynes and the Truckee Lumber Company, with the Southern Pacific and West Side Lumber Company sitting in, were destined to have the ultimate meeting regarding the future of Merrimac Forest.

The home of W. H. Kruger, still standing in Truckee, was built to overlook the mill to the south. Below: Site of the Truckee Lumber mill on the Truckee River, looking upstream. (Both photos Paul Beckstrom)

Swayne Lumber Company No. 4 Shay switching the original mill at Pearl Orchard, California. (Mallory Hope Ferrell collection) At right: Robert H. Swayne, partner and shareholder in Swayne Lumber Company and Swayne & Hoyt, Inc., from a snapshot taken aboard ship in the summer of 1932. (Courtesy the estate of Lloyd Swayne)

CHAPTER TWO

Swayne Hill and Chino Creek Operations

While the Truckee Lumber Company was organizing its Oroville operations and building the Butte & Plumas Railway, another logging operation was underway just north of an area later named Swayne Hill. This location, only one and one-half miles as the crow flies from Truckee's ill-fated French Creek grade, operated concurrently with the Truckee Lumber Company efforts to harvest its timber holdings in the Merrimac district. The corporate name for this adjacent (but not affilliated) enterprise was the Swayne Lumber Company.

In June 1909 Robert H. Swayne of San Francisco, his son Warren and Major Charles Tilden and his son Lloyd, formed the Feather River Lumber Company.[1] The elder Swayne was a partner in the San Francisco firm of Swayne and Hoyt, a customs house and brokerage firm. Swayne was also involved as a principal in the Albion and Southeastern Railroad (on California's redwood coast) from April 1902 until it was purchased by the Santa Fe Railway in January of 1905.[2] Swayne also was the owner of the Albion River Steamship Company which had three steam schooners, *Celia, Albion River* and *Pomo*. He apparently

SWAYNE AND HOYT
The Maritime Empire of Robert H. Swayne

The little steam schooner POMO,* pictured here some time in 1905 about to dock at San Francisco's Mission Street wharf, was but a small part of the Swayne and Hoyt shipping lines. In part, these steamship lines reflected the desire of lumbering concerns (those who could afford it) to vertically integrate their resources and assets. Swayne was well aware of the fact that, if he could control the transportation and distribution of his lumber products, his position in the market would be much stronger than many of his competitors.

The complete story of Swayne and Hoyt is beyond the scope of this book. We can, however, highlight the significant dates, firms and vessels that made up the seagoing side of the Swayne family enterprises.

The Edson family founded the predecessor company in 1850. The partnership was formed in 1896, between Robert H. Swayne and J. G. Hoyt. They were involved in their first lumbering venture as early as 1902 when they obtained the Albion Lumber Company and the Albion River Railroad. By purchasing the railroad, they got the log and lumber haul, and by providing for steamers, they controlled the shipments of lumber to San Francisco. They incorporated a separate firm, the Albion River Steamship Company, and equipped it with three wooden steam schooners, CELIA, ALBION RIVER and POMO.

Having got their feet wet, so to speak, the partners realized that, beyond operating a few small schooners out of the dog holes on California's Mendocino Coast, a vast opportunity awaited those who would venture into Pacific coastal and intercoastal shipping activities. In quick succession, the partners began to procure existing companies or open new routes with acquired or newly built vessels:

1906: Albion Lumber Company, with wood steam schooner PASADENA.

1909: With the Tilden family (mentioned in the text), started the California & North Pacific Steamship Company (later renamed Western Steam Navigation Company) with three vessels, including CAMINO and NAVAJO, running from Long Beach to Portland, Oregon.

1912: Long Beach Steamship Company, with steel steamers ALVARADO and PARAISO.

1913: Yellowstone Navigation Company, with wood steamer YELLOWSTONE.

1913: Arrow Line, with four vessels, for service from San Francisco to Portland.

1917: Western Transportation Company with steel steamer ELDORADO.

San Francisco Maritime National Historical Park Collection

All these firms were, of course, coastal and within United States waters. By 1917 the partners were involved in international trade in a big way. They landed a mail and shipping contract with the Mexican government and operated four more vessels under the Gulf Mail Steamship Company flag. In time these contracts led to two other flags: Gulf Pacific Lines (intercoastal from Tacoma to San Francisco and Long Beach then via the Panama Canal to Texas, Louisiana and Florida) and The Gulf Pacific Mail Line, truly international in scope, operating from the Pacific Coast via the Panama Canal to Colombia, Venezuela and the Mexican ports of Veracruz and Tampico, as well as Jamaica and the Dutch West Indies.

During the Depression of the 1930's, some additional vessels were purchased by Swayne and Hoyt for the Caribbean service and were upgraded with passenger accommodations. In a salute to the firm's original area of operations, some of the vessels were renamed for prominent headlands to the north and south of San Francisco: POINT ARENA, POINT LOBOS and POINT REYES.

In April 1940, coinciding with the closing of Swayne Lumber Company and its narrow gauge railroad, Swayne and Hoyt ended their maritime operations when the last vessels were sold.

—S&H data courtesy Kramer A. Adams

*The POMO was built in 1903 for S&H's Albion River Steamship Company by the Bendixsen Shipbuilding Company at Fairhaven, California. She was a typical wooden steam lumber schooner of 368 tons, equipped with a triple expansion engine built by Montague Iron Works, and could carry 300,000 board feet of cut lumber. The POMO's end came on December 31, 1913 when she foundered in a storm off Point Reyes.

kept ownership of this latter firm until the *Pomo* was lost off Point Reyes in December 1913.[3]

Swayne's partner, John G. Hoyt, had formed Swayne and Hoyt, Inc. in 1896. This coastal shipping line served the west coast, as well as foreign ports, dating back to the redwood lumber hauls of the 1850's. Major Tilden was also an attorney and a corporate executive of some note.

The Swaynes soon discovered that the Feather River Lumber Company had the same name as an outfit running out of Delleker, further up the canyon; hence the name was changed to Swayne Lumber Company to avoid conflict. With the arrival of the Feather River Railway in the same general area in 1940, all of the different "Feather River" railroads have since served to confuse historians.

Swayne and Tilden purchased 2600 acres of timberland in the Chino Creek drainage from H. H. Yard for $180,000. Yard had worked as an agent for the Western Pacific Railroad in securing their right-of-way through the canyon, and had the foresight to acquire timber holdings at the same time. The lands purchased were sections 9, 10, 15, 16 and 22 of Township 22N, Range 5E.[4]

A site for the mill was determined and land purchased for this purpose in the southwest quarter of Section 27, and the northeast quarter of the northwest quarter of Section 34 (T22N,R5E).[5] This area today is known as Swayne Hill. At the time of mill construction, it was referred to as the Pearl Orchard/Camp Enjoyment area. The Swayne millsite was located just north of Camp Enjoyment, a Feather River canyon resort of the era, approximately 25 miles northeast of Oroville, California.

Camp Enjoyment was incorporated in 1909 with visions of becoming a permanent summer community. The camp was billed as a mountain resort, a "quick" forty minutes by the Western Pacific Railroad up the Feather River from Oroville.[6] Camp Enjoyment offered a two-story hotel called "El Sierra," tennis courts, a lake called "Perla Laguna," 280 40-by-80-foot lots for sale, croquet courts and an outstanding view of the Feather River canyon. The short lifespan of Camp Enjoyment can no doubt be attributed to its proximity to the Swayne mill and drying yards.

By August 1909 plans had been finalized for the mill and work was underway in preparation for the erection of the mill buildings. Horses were sent from San Francisco for use in clearing the millsite.

The Swayne Mill was completed between August 1909 and July of 1910. The complex featured a 45-by-120-foot sawmill, a planing mill, a box factory, drying kilns, bunkhouses, a cookhouse and housing for married employees. The mill featured a nine-foot band saw with a five-saw gang edger. Power was furnished by two large Dutch Oven boilers set on concrete pads and enclosed by brickwork. The boilers burned sawdust piped from the saws as well as other mill debris and waste which was first run through a "hog" that cut it into small pieces.

Extensive drying yards surrounded the mill. These yards were served by three distinct narrow gauge tram systems. At full production, the mill and its logging operations employed 150 men.

Warren H. Swayne, president of the lumber firm, had a large home built on the hill overlooking the mill. Locals referred to the structure as the "Swayne Mansion."[7] The Swayne Hill name resulted from the construction of the home. The "mansion" had running water, piped from the Chino Creek ditch, and an electrical generating system shared with the mill. Ray Hughes purchased the Swayne home in 1922, and lived in it until the 1960's, when a forest fire burned through the area and destroyed it.

The Swayne mill was connected to timber cutting areas to the north by three miles of 36-inch gauge railroad. Logs were yarded and loaded onto railroad cars with a steam donkey. Rolling stock was a mixture of second-hand flatcars and wooden disconnected trucks. Motive power in 1910 was a Class A Climax. In 1911 a two-truck Shay, c/n 2458, was purchased new and supplemented the Climax until 1916.

The most unique feature of the Swayne Hill operation was how finished lumber reached the market. The closest access to interstate transportation was the Western Pacific Railroad, passing less than a mile to the southwest, but 4200 feet below in the bottom of the Feather River canyon.

Swayne's solution to moving lumber from the mill to the Western Pacific line was to build a tramway down the hillside to the WP near the mouth of French Creek. This was the same siding where the Truckee Lumber Company had proposed to begin its French Creek grade in 1909. The station point was "Blinzig," named after a German financier who had toured the Western Pacific with owner Gould, remarking that people didn't have to travel as far as Switzerland to see grand scenery.[8]

The Swayne tram was unique in that it was the first tramway built in Butte County. Powered by a forty-horsepower engine, it had a twelve-foot diameter cable drum[9] and its trackage was 4200 feet in length. A half-mile of narrow gauge trackage connected the tram head with the Swayne mill.

The tram was finished in May of 1910 and was utilized to haul machinery to the new mill. In mid-August of that year the mill was completed and began sawing lumber. The first carload of lumber was sent down the tram on August 29, 1910.[10]

After completion of the Swayne tramway, Western Pacific extended its spur at Blinzig, establishing a loading track called Swayne Siding. A section house, water tank,

Camp Enjoyment, May 1909, with Swayne Railway indicated at upper right. (Courtesy Butte County Recorder's Office)

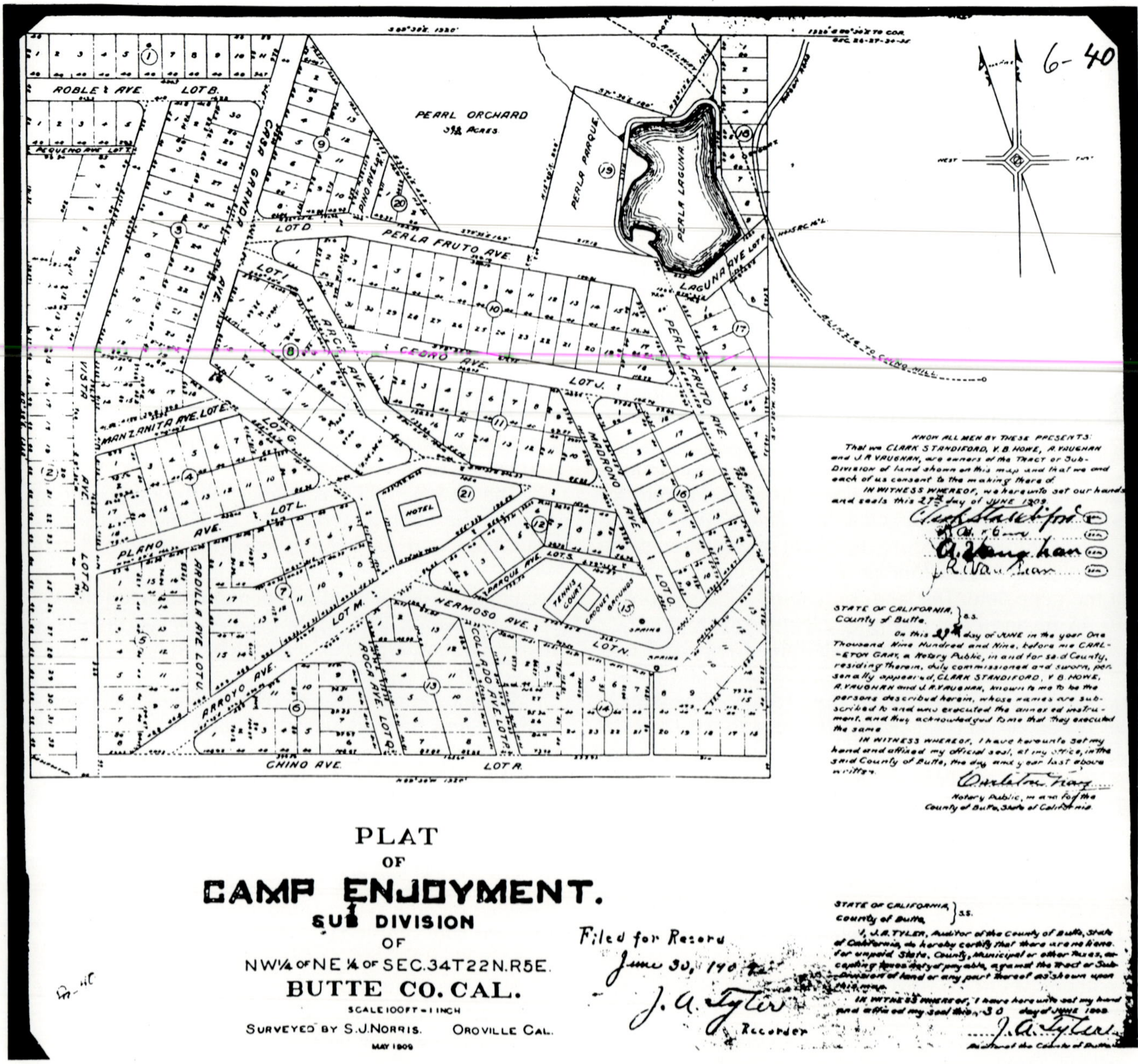

toolhouse and other ancillary buildings were erected along this siding. A station was also built at Blinzig to accommodate passengers traveling to and from Camp Enjoyment, Stanwood and French Creek. At this time, a steep trail linked Camp Enjoyment with Blinzig, as the Swayne tram did not routinely carry passengers other than tram workers.

In an interesting enterprise, contemporary with the Swayne tram, was the construction of yet another tramway by a man named Clark Standiford. Standiford was the owner of the Pearl Orchard Ranch and chief promoter of Camp Enjoyment. He also owned timber holdings just east of Swayne's in the Chino Creek Basin.[11]

In February 1911 the Butte County Land and Lumber Company was incorporated with a capital stock of $75,000. Standiford and investors from Pasadena, California, where Standiford spent his winters, were principals in the organization. Standiford, who served as a director, then sold his holdings in Section 22 to the new company.[12]

The Butte County millsite was just south of the Swayne mill in Section 34. Prior to construction of the mill, Standiford convinced the company to construct a tram parallel to and south of the Swayne tramway. In fact, he maintained that the tramway down Swayne Hill to Blinzig Siding was his idea all along and that Swayne had "borrowed" the plan in building his first.[13]

The Butte County tramway, completed in June 1911,

Mr. Hoyt and Mr. Swayne Sr., boarding the narrow gauge for a ride to the tramway, Swayne Hill area, 1914. Locomotive is the class "A" Climax. (Authors' collection)

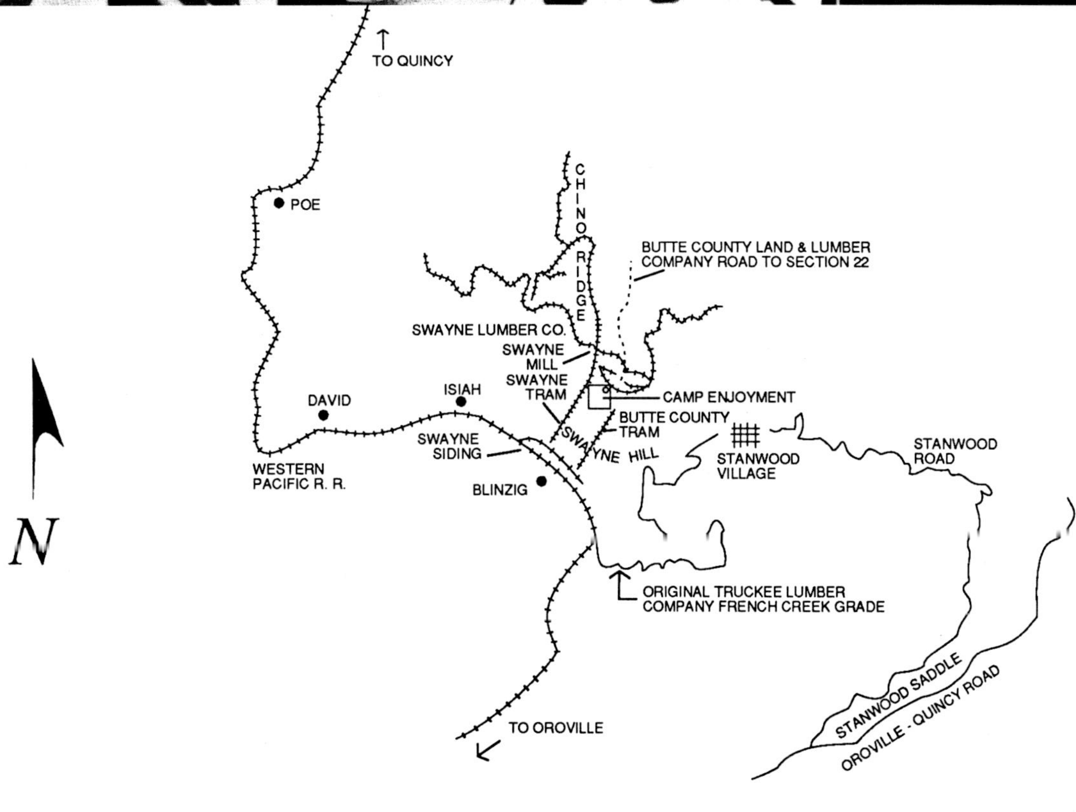

Map by Bob Mariano

SWAYNE HILL OPERATIONS (1909 - 1917)
SWAYNE LUMBER COMPANY

was described earlier in the May 27th edition of the *Oroville Daily Register* as follows:

> "The incline tram railway being built by the Butte County Mill and Lumber Company to connect Camp Enjoyment and the company's mill and lumber yard with the main line of the Western Pacific Railway is nearing completion. The grading has been finished all the way down the side of the mountain and ties are being laid. The work will be completed by June 1st.
>
> "Sawed cedar ties are being used, with red fir stringers, 3×6 inches, spiked to them, on which the rails will be fastened with lag screws and spikes, the rails being bored for the screws. These screws will keep the rails from creeping down the moutainside.
>
> "The tram will be 3700 feet long and makes a perpendicular rise of 1800 feet in that distance to reach the elevation of the lumber yards and Camp Enjoyment. The novel feature of the railroad is its steepness, making a shortcut from the bottom of the canyon to the top of the mountain.
>
> "Cars will be run on this tram to meet all Western Pacific trains and will carry passengers, baggage, mail, express and freight. It will make Camp Enjoyment easy of access."[14]

A further description appeared in the *Register* on July 25th:

> "The tram is known as a three rail, balanced tram and is 3800 feet long and makes a vertical rise of 1800 feet in this distance. The track is straight from top to bottom and cost about $12,000. The grade varies from 20 to 90 degrees.
>
> "The materials used in construction are cedar ties with red fir stringers spiked to them. The rails are laid on the stringers. This insures the track being well tied together.
>
> "A car can make the run in seven minutes and as a car ascends, another car will descend. The hoisting machinery was especially made for this purpose and has several brakes, any one of which is capable of handling the loads. A seven-eighths steel cable will be used of the best make which has a pulling capacity of more than 50,000 pounds on this tram.
>
> "Clark Standiford first conceived of the idea of this tram for the hill between Blinzig on the Western Pacific and the country around Camp Enjoyment — and the Swayne Lumber Company installed one on similar lines to the one built by the Butte County Land and Lumber. Mr. Standiford located and planned the details of this tram, and it was built by James and Edgecomb."[15]

The Butte County Land and Lumber Company mill was completed a year later in July 1912. The mill had a potential capacity of forty thousand feet daily. Timbers cut by the mill were utilized to build a blacksmith shop, a boilerhouse and various outbuildings.

The Butte County Land and Lumber Company timber holdings comprised the eastern half of Section 22, Township 22N, Range 5E. In order to reach these lands and transport logs back to their mill in Section 34, Standiford built a logging and wagon road through Section 27, the majority of which was owned by Swayne. Standiford used a steam traction engine to move logs along this road which passed near Swayne's mill and drying yard on its way down to the Butte mill.

The fact that Butte County Land and Lumber trespassed on Swayne land in building its new road, and posed a fire hazard to the Swayne mill due to the passing of its traction engine, prompted Swayne to file injunction proceedings. Swayne Lumber Company charged that its lands were entered by Butte County Land and Lumber Company on August 1, 1912 without permission and over Swayne's protest, and a logging road subsequently constructed. Five hundred dollars in damages were requested. The injunction further stated the operation of the traction engine posed a fire hazard to the Swayne's mill.[16]

The final blow was delivered to Standiford's company late in 1913 when a decree of foreclosure was handed down in favor of the Bank of Rideout-Smith and Company for a $3000 note. Butte County Land and Lumber Company was now bankrupt. They sold the mill to the Quincy Lumber Company and the machinery was later dismantled and shipped to Quincy.[17] The tramway was also dismantled and Swayne consequently dropped its lawsuit.

From 1913 to 1916 Swayne Lumber Company continued to log the forests surrounding the mill, purchasing additional timberlands from the Central Pacific Railway. These holdings, in Sections 5, 17, 19, 21, 23, 27 and 29, Township 22N, Range 5E), accounted for an additional four thousand acres[18] and cost $42,840.

The Swayne mill ran spring, summer and fall, closing down during the winter. Most of the mill workers lived nearby in the hamlet of Tilden, named after Charles Tilden, the Vice President of the company. Other workers lived at Stanwood, a mile and a half to the south. To improve communication with the Sacramento Valley, a telephone line was run from Berry Creek to Tilden.

In 1914 "Tilden" was changed to "Swayne." Warren Swayne became the new postmaster and the town endured until June 1917. While it lasted, Swayne boasted tri-weekly mail service and a direct stage line from Oroville.

The year 1916 marked a troubled year for Swayne Lumber Company. On September 11th a forest fire started on Big Bar Mountain. Before it could be contained, two hundred acres of prime timberland was destroyed. Ninety Swayne employees fought the fire, successfully preventing it from reaching the town or the mill. Six weeks later a fire of suspicious origin destroyed Swayne's mill, drying kilns and the entire year's output of lumber. Firefighters managed to save the planing mill, box factory and nearby homes, but the loss, some of which was covered by insurance, was reported at $300,000.

The October 26th issue of the *Oroville Daily Register* reported the following:

> "The Swayne Lumber Company, one of the largest lumbering establishments in the Feather River Canyon was destroyed by fire at an early morning hour yesterday.
>
> "The entire season's lumber output was burned, together with the mill and dry kiln. It is estimated that six million feet of lumber were destroyed in the conflagration. The planing mill and box factory, both small and unimportant buildings, and the cottages in which the men lived, alone were saved. No statement would be made by officials of the company as to the loss, but those familiar with the plant here, estimated the loss to be in the neighborhood of $300,000. It was partially covered by insurance.
>
> "The fire was first detected by the night watchman shortly after midnight. It broke out in the lumber yard beneath one of the piles.
>
> "The flames spread with incredible fury. The whole camp responded to the clanging of the fire bell, and streams of water were soon upon the fire. Sixty men were engaged in fighting the flames. They climbed upon the lumber piles, and a number of the men narrowly escaped death when a pile on which they were standing to fight the flames, collapsed.
>
> "When the flames reached that portion of the yard where the dryer timber was stacked, the heat became unbearable. The whole heavens were lighted with the flames of the fire. The timber burned with a white heat that drove the men back from the flames. The fire jumped from one stack to another.
>
> "In a vain attempt to save the mill, dynamite was used and an attempt was made to clear a path between the yard and the mill that the flames would not cross. The effort proved futile. The mill was soon on fire, and the dry kiln burned next.
>
> "Driven back by the intense heat, the men were forced to abandon their fight to save the mill and lumber property and devote themselves to saving their homes. All night long the conflict continued. That the cottages were saved was due almost wholly to the protection afforded them by the trees.
>
> "The Company had a very elaborate system of caring for its lumber in its yard, involving three distinct tramway systems. All of this was wiped out by the flames.
>
> "It is estimated that from 100 to 150 men will be thrown out of employment by the fire. The woods camp was closed yesterday and in a few days only the watchman and a few others will be left.
>
> "No statement could be made yesterday by the officials upon the ground as to the cause of the fire or the future plans of the company. The main offices of the company are in San Francisco, and it was stated that such information would come at a proper time from the main offices. The company has a large forest of fine timber, and it is thought probable that the mill will be rebuilt for next year's run.
>
> "Yesterday any attempt to get at the company books or to settle with the men for their labors was frustrated by the fact that Bookkeeper Miller, some time during the course of the firefighting lost his keys to the safe. It is probable that the safe will have to be blown in order to pay the men off."[19]

The big fire of October 1916 marked the end of decentralized lumber milling in the Oroville area woods. From the 1850's to the date of the fire, typical lumbering operations were small, independent outfits, located close to the timber to be cut and supplying lumber within their own local area. The operation which would replace the ashes of the Swayne Hill mill was destined to be markedly different and on a much larger scale.

Log dump at Swayne mill near Chino Creek in 1914. Locomotive is Swayne No 2/4 burning wood. Flatcar at left is an ex-NWP 28-footer built by Carter Brothers. Cars were cabled onto the light trestle. (J. E. Boynton collection)

Above left: The lumber-sorting area of the Swayne Hill mill, about 1914. Below: Another view of the mill under a coating of light snow. (Both photos J. E. Boynton collection)

The mill at Oroville in 1918. Locomotive is Shay No. 1. (California State Railroad Museum)

CHAPTER THREE

The Swayne Lumber Company In Its Prime: The Twenties

The big Swayne Hill mill fire of October 1916 forced the Swayne family to consider several options for the future. One choice would have been to shut down the operation and sell the timberlands. Another would have been to rebuild the mill and continue as before. A third option — the one finally taken — was the most ambitious and daring: to negotiate the purchase of the Truckee Lumber Company steel mill in Oroville and the attendant Butte & Plumas Railway. Both properties had sat idle since October 1911, despite efforts of the various owners and lienholders to dispose of them (as outlined in Chapter One).

The Swaynes obviously gave a great deal of thought to taking the company in this new direction. Instead of being one of the small sawmill operations in and about the Feather River canyon area, the new outfit would be one of the largest in terms of equipment, trackage and cutting area. As the dollar amounts involved were beyond the Swaynes' credit line, they knew they would have to negotiate hard with the real owner of the property, Southern Pacific. If the SP decided to go with the Swaynes, then the other two parties — Truckee Lumber Company and West Side Lumber Company — would have little choice but to go along to salvage anything of their original investments.

That is exactly what happened when an overall agreement for the entire property was reached in April of 1917. The Swaynes represented the only truly serious buyer that Southern Pacific had seen in almost six years and this fact probably worked as a trump-card in the Swayne hand. As the Truckee Lumber Company had defaulted on the original $400,000 note to the Southern Pacific for the Butte & Plumas Railway construction, the railroad was the SP's to sell. The Truckee Lumber mill was still in the name of Truckee, although bound by liens from, among others, West Side Lumber Company. The timberlands, also Truckee property, were under similar liens.

The Swaynes got the railroad for a $250,000 note, which meant the SP wrote off $150,000 on the Butte & Plumas. The cost of the Truckee mill is unknown. West Side Lumber reduced the $1 million note to $750,000, payable over ten years, releasing the timberlands to Swayne.[1] What is not clear about the West Side's reduced note is exactly who was to pay. An educated guess would place Truckee Lumber Company as the primary obligator, with Swayne paying Truckee for leased timberlands, and Truckee in turn paying West Side in increments. Such was the best solution for Swayne and

A panoramic view of the Swayne mill in 1917, looking east. Note the caboose on the tail of the wye, slash pile burning behind the office, and the dredger tailing piles. (Ed Thomas collection)

Truckee, but not for West Side. As Swayne did not want to purchase the timberlands, but only harvest them, being a sublessee to West Side through Truckee, was definitely in Swayne's favor. Whether Truckee actually paid West Side was of very little concern to the former and, as we have already seen, by 1924 West Side was largely left holding a bad note while the Swayne Lumber Company had profited in the meantime. A timber cruise report from the 1920's shows Truckee Lumber Company still a majority landholder in the Merrimac Forest, which would seem to confirm the actual situation.

As Truckee was a bad debtor to West Side Lumber, so ultimately was Swayne to the Southern Pacific. The $250,000 note for the railroad was never fully paid and, as will be seen later on, the Southern Pacific once again had to repossess their narrow gauge railroad in 1939.[2]

Over the life of the company from 1917 to 1939, the Swaynes had a railroad and property that was either only partially paid for or covered by lease. The wonder is in all that time that neither the Southern Pacific nor West Side Lumber Company went to court to recover what was in essence their property. Be that as it may, after the Spring of 1917, no further notice of various lienholders was to appear until the road closed in 1939.

So far as is known, no train movements took place on Butte & Plumas trackage from October 1911 through the Spring of 1917. The seriously deteriorated trackage was largely rebuilt, as were most of the 64 trestles and the big bridge over the Feather River. The El Oro Dredging Company had cut the line near Oroville in a minor revival of gold dredging.[3] Due to the high gold content of the Oroville district gravels, that town had seen gold dredging activity since 1898. In fact, more gold dredges were floated in Oroville than in any area in California.[4] The entire Truckee mill and drying yard had been constructed on leveled dredge tailings and the dredgings of the 'teens represented the final reworking of the Dry Creek gravels. Swayne Lumber Company replaced the displaced roadbed in the tailings, regrading and relaying the tracks.

As a final reminder of the Truckee Lumber Company/Western Pacific disagreements, the Swaynes had to negotiate with the WP for a right-of-way for the newly rebuilt narrow gauge under the WP main line through Oroville. There was little choice, since the WP tracks were located just east of the mill property and the route was the only way east to the woods. In granting this right-of-way under its tracks, Western Pacific probably

had no idea that Southern Pacific would install a third rail on the narrow gauge to gain its own access to the Swayne mill grounds and that this rail would remain in place until the end. The Northern Electric, working with the Western Pacific, countered with an extension of NE trackage, past the Union Ice plant and onto the mill property. This extension was under trolley wire and was probably the only instance of a narrow gauge logger being switched by an interurban. The NE was reorganized as the Sacramento Northern in 1918. Some time after 1926, the SN trackage which to that point merely crossed over Swayne trackage with no dual-gauging was three-railed next to the power house.

By the summer of 1917 the rebuilding and expansion efforts of the Swaynes merited the attention of *The Timberman,* as witness their June issue:

> "OROVILLE — The Swayne Lumber Co., successors to the Truckee Lumber Co., is busy getting the sawmill in shape for operation. The mill is a modern Filer & Stowell two-band mill of steel construction throughout. The power is generated by two Sterling water tube boilers. The power house is of steel, brick and concrete construction. The power is derived from 28″×48″ Corliss engines developing 750 horsepower. The carriages are equipped with shotgun feeds; two edgers and a complete system of live rolls and transfers, trimmers, slashers and lath mill. On the main sawing sides, a Simondson rigger is used with a Hill rigger on the pony side. A drag sawmill will be installed on the log deck and the pond for cutting up logs and facilitating operations.
>
> "The mill was built about six years ago and operated only about five months, when it was closed. Repairs are being made and additional machinery is being installed in the planing mill, which is electrically driven, part of the equipment being removed from the company mill formerly located at Swayne, on the Western Pacific, which was destroyed by fire. The dry kilns are of the standard type and of concrete construction. A box factory will be added later. A Holt tractor has been purchased. The timber holdings consist of about 600,000,000 feet of excellent timber of which approximately 65 per cent is pine. The Butte & Plumas narrow gauge railroad, 23 miles long, was taken over by the Swayne Lumber Co. Ten miles of additional line will be built at the terminus of the present road, the steel having been previously

Merrimac camp in 1919. No. 3 is on the left with a swing from Stanwood Saddle. No. 4, at right, headed back to Gold Lake Camp. Ford speeder belonged to Superintendent Pat Lyons. Note the early riveted bunks on the log cars. (Paul Beckstrom collection)

employed on the Swayne Lumber Co.'s road at Swayne, where are the timber holdings of the Swayne Lumber Co., with the Truckee Lumber Co.'s former holdings adjoining. The locomotive equipment consists of Shays and four direct connected locomotives. A new 60-ton Shay locomotive has just been purchased. The logging cars are of the Pacific Car and Foundry Co.'s standard type equipped with air. The mill will have a 10-hour capacity of about 120,000 feet and will be ready to operate probably in July.

"The men who will assist President and Manager W. H. Swayne in the operation of this modern plant are: Superintendent, J. E. Ring; filer, R. W. Wright; logging superintendent, Harry White. In the starting up of the mill and railroad, R. H. Swayne, father of Manager Swayne and an active member of the well-known lumber and shipping firm of Swayne and Hoyt, San Francisco is giving all the time he can spare. The Swayne Lumber Co. will add to lumber activities of the Feather River Canyon and will be a helpful factor in the trade. The town of Oroville will be greatly stimulated through the operation of the mill. The Southern Pacific is putting a spur into the mill. The Western Pacific will also make direct track connections, this company reaching the mill at present over the tracks of the Northern Electric Co."[5]

One snag occurred when it was time to fill the millpond. Truckee Lumber had purchased water from the Palermo Land and Water Company; however, the Railroad Commission, in the intervening years, had ruled that water could not go to mills if there was a need for the same water for irrigation. A solution was finally found by drilling deep wells on the mill property.[6]

On June 30, 1917 the newly rebuilt mill opened and began to cut logs hauled in by the railroad from Berry Creek. Things did not stay rosy for long. World War One had caused a severe shortage of loggers and mill hands. By now the limited timber holdings available to the railroad caused the mill to cut at only one-half capacity.

It became readily apparent that extension of the railroad was necessary to access the company's timber holdings in the Merrimac area. During late 1917 and early 1918 efforts were concentrated on extending the line to Merrimac. This portion of the line was fittingly called the "Merrimac Extension."[7] One hundred men, a steam shovel and horses with Fresno Scrapers were utilized in this construction effort. By mid-1918 the railroad had reached Stanwood Saddle and by late 1918 the site of Merrimac.[8]

This thirteen-mile extension was significant because it opened up the vast Merrimac Forest to railroad logging. The small town of Merrimac was itself to become an important railroad junction in later years.

The many circuitous logging spurs that served this area for the next twenty years all fed back to Merrimac, and the run from Merrimac back to Oroville became the Swayne "mainline."

In order to provide adequate logs for the Oroville mill, several additional tracts of timber were purchased during 1917 and 1918. Swayne purchased all of Section 19, Township 21N, Range 3E from the Central Pacific Railway for $19,108.63.[9] The northwest quarter of Section 7 and the western half of Section 5 Township 21N, Range 6E were purchased in 1918 for $10,163.83.[10] These purchases were evidently logged during 1918 and 1919.

In the 1920's the Swayne operation was of medium size in comparison to California giants — The Red River Lumber Company, McCloud Lumber Company, and Weed Lumber Company. Nevertheless, Swayne's production was much higher than the myriad of small mills in the state, and was one of the largest Feather River country mills. The lumber sheds and box factory were serviced by the Southern Pacific Oroville Branch, with an average of four to six boxcars shipped daily.

During the years 1917 to 1919, production was less than twenty million board feet per year. By 1920 production was raised to 23 million board feet and the Oroville box factory was running year-round.

By 1929 the height of the railroad logging era in the Merrimac forests was passing. The Twenties witnessed steam railroad and steam donkey logging involving several hundred men in the areas of Palmetto, Granite Basin and Sly Creek. Swayne's operations in the 1920'S were typical of most other sugar pine logging operations in the Sierra Nevada. The logging and railroad season was from April to mid-November. Although crews were laid off the rest of the year, a few lucky railroad workers were retained through the winter maintaining and rebuilding locomotives and rolling stock.

The Swayne's railroad operations were efficient and safety-conscious in many ways, employing such innovations as superheaters, feedwater systems, piston valves on locomotives, and straight and automatic airbrakes on the log cars (installed in the late 'teens). The Worthington feedwater systems no doubt reflected parent company Southern Pacific's preference for this manufacturer. A further indication of the company's efficiency was the presence of flange oilers at each end of most trestles. Employees remember both water and oil being used in these devices.

The main line locomotives were Shays 1, 2 and 6; the woods engines were 3 and 4 and Heisler No. 5. All locomotives were repainted each winter. Numbers, lettering and striping varied between silver and white, depending on the year in question. Photographs show well-maintained locomotives with little evidence of weathering. Locomotive firemen furthered this effort by wiping down the boiler jacket with a mixture of lampblack and coal oil as the trains made their way downhill between Merrimac and Bidwell Junction.[11]

A Swayne donkey near Merrimac in 1924 with three local Indians among the crew. (Paul Beckstrom collection)

Skeleton log cars were rebuilt as necessary with new timbers retrofitted and mechanical items replaced or rebuilt. Skeletons were painted oxide red on a hit-or-miss basis. Both photographs and employees' recollections indicate that some skeletons were painted entirely red, while some had only the ends painted; however, most were left unpainted. The skeleton cars of the nearby Hutchinson Lumber Company, interestingly enough, were also unpainted for the most part.

While rolling stock warranted yearly attention, the trackwork was a lesser priority. Ex-employees remember trackage, including the main line, to be in poor shape with light rail and dirt ballast. Track was "soft" in many places, resulting in frequent derailments.

No. 5 Heisler was the woods switcher at Baker Creek. Truck/rail reloads eliminated this function in the late 1930's. Woods switchers did not exhibit Swayne's usual high maintenance since they stayed in the woods all season long. (E. E. Slaughter)

No. 4 is also at Baker Creek in the late 1920's. This was a crew change point and here, with a picture taken from the empty haul, a load of logs is ready to depart for Oroville. (J. E. Boynton collection)

Severe winter weather undoubtedly played a major role in track condition, as adequate section crews were provided. Section crews, each consisting of four or five men and a foreman, were stationed at Oroville, Bidwell Junction, Berry Creek and Brush Creek. A bridge crew was also stationed at Bidwell Junction and was responsible for all bridge, trestle and culvert maintenance. In addition, a steel gang of twenty men was kept busy constructing new logging spurs and tearing up old ones. The steel gang foreman was an ex-San Francisco policeman by the name of Matthew O'Leary, who was reported to be as strong as three men.[12]

Logging operations in the 1920's consisted of axes and crosscut saws, horses, steam donkey engines, high lead cable systems and logging camps. Swayne owned up to fourteen steam donkeys, the number varying from 1920 to the early 1930's. Each donkey was woodburning to the end and numbered according to its age, the oldest having the lowest number. Donkeys 1 and 2 were Dolbeers and were used only for pumping water to the other donkeys. The bigger donkeys were all built by Willamette Iron Works of Portland, Oregon, with the exception of No. 12, which was made in Tacoma, Washington by Puget Sound Iron & Steel Works.[13]

Horses played an important role in woods operations during the Teens and Twenties. Swayne maintained twenty to thirty horses in a corral in Oroville where a K-Mart store is located today. The animals were used extensively in constructing new roadbed with Fresno Scrapers, and also in logging operations. The horses were moved long distances over the railroad in a boxcar from which the roof had been removed. As some of the older donkey engines lacked haulback drums, the horses were also used to take the cable back out for each yarding movement. Additionally, cordwood for the donkeys was provided by two wood choppers and distributed to the machines by a teamster in a horsedrawn "wood" wagon.[14]

The hostler who cared for the horses was a Negro by the name of Charles Blackwell. He was an excellent employee and was well respected by Swayne management. Blackwell once lost several months' employment due to a broken leg caused by the kick of an unappreciative horse.

In conjunction with its steam donkeys and horses, Swayne also utilized high lead cable system technology. High leads were introduced to western pine operations during World War One, but it wasn't until 1923 that Swayne converted to high line operation from the common practice of yarding or dragging logs along the ground.

In a typical Swayne high lead operation, a spar tree was prepared and rigged with a main line and haulback line going through blocks set high on the spar. With this system, each donkey could move a log approximately seven to eight hundred feet. Sometimes relay systems were utilized, involving two donkeys working in unison, bringing in logs from distances of 1600 feet. Logs were yarded to landings and subsequently loaded on log cars by various methods ranging from tight or crotch lines to simple stationary block setups. *The Timberman*, in September 1923, described Swayne's operations as follows:

> "The camp at Palmetto is being operated with four sides. The system used is high lead, without a swing, and the loading is done with a separate engine with a block on one of the guy lines. One side is working on a National Forest timber sale and is using a modified lead of 35 feet. The logs are yarded in 32 foot lengths and are transported on connected trucks. The logging railroad is about 45 miles in length."[15]

The route of the Swayne Lumber Company railroad, known locally as the Swayne Railroad or more commonly as "Swayne's " began at the mill (now the site of Hobbie Chevrolet, 1250 Oro Dam Boulevard, in Oroville, California). The tracks continued to the east, ducking under the Western Pacific main line and the Southern Pacific Oroville Branch. The track then hugged the north bank of Dry Creek, crossing Meyers Street and eventually Dry Creek itself, to head uphill along the present alignment of Olive Highway. The railroad continued uphill through the orchard and vineyard of Pete Tufoni, and the olive orchard and meadow owned by George Wallace.[16] This location was known to the train crews as "Whitewashed Trees."

After crossing the Oroville-Quincy Road and the Palermo Canal (at today's Oakvale Avenue and Olive Highway), the tracks followed the canal for a short distance and then crossed at 51 Ward Avenue, swinging to the north behind the homes on Bessie Lane. Summit Siding was located at the end of Bessie Lane and the earthwork is still very visible. From Summit Siding to Bidwell Junction, just east of the crossing of the Middle Fork of the Feather River, was three miles of 4½-percent downhill grade. The present-day bridle/footpath at the southeast corner of Oroville Dam follows this steep section of railroad right-of-way. This grade was an operating factor for mill-bound trains (westbound; geographical south) in that coming up from Bidwell necessitated breaking trains into two or three sections in order to double or triple the hill.[17]

The Middle Fork of the Feather River was crossed on a high bridge about three hundred yards upstream from the Western Pacific's crossing. Just east of Swayne's bridge was Bidwell Junction. A siding, water tank and several cabins were located here with a section gang and bridge gang operating from this point. Bidwell Junction was two miles downstream from Bidwell Bar, now under the deep waters of Lake Oroville. Bidwell Bar was the site of California's first suspension bridge and the home of General Bidwell's famous "Mother Orange" tree. As there was no road into Bidwell Junction, workers parked their automobiles at the Simmons Ranch on the Oroville-Quincy Road and walked in.[18]

Downstream from Bidwell Junction and on the opposite bank of the Middle Fork was the Western Pacific station of Land, MP 212.2, seven miles from Oroville. Land was unique in that it occupied a location where the rails of the Western Pacific, Swayne Lumber Company and the Hutchinson Lumber Company (later renamed Feather River Pine Mills, and operating the railroad after 1940 as "The Feather River Railway") all came within three hundred yards of each other. The name came from A. H. Land, who was an original backer of Hutchinson, and later president of the successor Feather River Pine Mills Company. Hutchinson crossed Swayne's track at grade, with a derail installed uphill of the Swayne track on the Hutchinson standard gauge line. Although Swayne trains would have right-of-way by virtue of the B&P tracks being laid first, the management quickly made it mandatory for east and westbound movements to come to a complete stop when reaching the crossing. After checking the Hutchinson track, the Swayne train would proceed, with the crossing being flagged. After awhile the safety precautions became somewhat of a running joke among the crews of the 4:20 and 9:00 a.m. MTy turns, since one employee observed, "the Hutchinson crew generally didn't get going much before 11 a.m.!"[19] By that time of day at least three Swayne trains would have already passed through the crossing.

From Bidwell Junction the railroad continued uphill through the Canyon Creek drainage, passing railroad landmarks of Rogers' Crossing and the long passing siding at "42 " Halfway up Canyon Creek was the water tank at Blacks Crossing. Cynthia Black, a widow, was paid a small monthly wage for making certain the tank was full.[20] In order to accomplish this task, she operated a manual hydraulic ram pump that lifted water all the way up from Canyon Creek.

The next water tank was at Hartsmill, as were the railroad's two highest trestles — No. 53, the Highway Bridge which crossed the Oroville-Quincy Road, and No. 54, the Hartsmill Trestle. The latter was 112 feet high and employees confirmed a wager that the trestle swayed thirteen inches from side to side as trains passed over it.[21] The Hartsmill tank sat on a knoll between the two trestles as did a small cabin. A man named Dow Day was paid to keep this tank full and to check the trestles for fire as each train passed. The tank was filled by a one-cylinder distillate pump placed in the creek. A pipe ran from the tank along the top of the trestle and a length of 1½-inch fire hose was provided for fire protection.[22]

Just east of Hartsmill, the Zink Branch left the main line, generally following the route of today's Bald Rock Road and terminating about two miles near the Zink Ranch.

Arid Circle, directly across the road from the Berry Creek Post Office, was the site of the next section house and crew, uphill from Bidwell Junction and the last one until Mosquito Creek. From there the railroad ran almost due east to Lake Madrone. Prior to the building of Lake Madrone Dam in the 1930's, the railroad ran along the south side of Berry Creek and crossed it about halfway up the present lake site. Once the dam was constructed, the line was relocated below the face of the dam and to the north bank of the lake. The Berry Creek water tank sat on the south bank of the lake where the Lake Madrone Inn stands today. One mile above Lake Madrone was a white stake marker that read "end of Southern Pacific property."[23] The marker served to remind even the casual observer of the real owner of the former Butte & Plumas Railway.

The Swayne railroad crossed the Oroville-Quincy Road near Berry Creek Cemetery and continued uphill in a northerly fashion to Stanwood Saddle, a prominent topographical feature and site of a passing siding. A mile above Stanwood Saddle was Mosquito Creek and Mosquito Junction. In the Twenties and early Thirties, Mosquito Junction's spur and oil facility served as an important meeting point for trains trading loads and empties.

Beyond Mosquito Junction were the important logging areas of Four Mile Ridge and Sly Creek. Four Mile Ridge was the site of many short spur logging lines. Six miles above Mosquito Junction was the small town of Merrimac and to the northwest was Merrimac Camp, or just "Merrimac," which was the site of yet another water tank. Additional tanks existed at Gold Lake Camp, Baker Creek, Granite Basin and at most other camps as well.[24]

When operating in the woods beyond the last water tank, train crews used the locomotive siphon to fill the engine tank from nearby creeks.[25] A bucket would be placed in the water and the siphon hose placed in the bucket to prevent siphoning mud or gravel. Locomotive sand was located both at Oroville and at the camps where it was kept in 10-20 barley sacks. When caught low on sand in the woods, crews would use decomposed granite indigenous to the area.

Oil for the engines was located at Oroville, at Mosquito Junction and at the camps. Most camp oil facilities were merely a tank car spotted on a spur. These cars were spotted so they would stay in the sun most of the day, heating the heavy Bunker C oil inside, and allowing it to flow through a hose by gravity.[26]

Three miles above Merrimac Camp, at Haphazard Creek, was Gold Lake Camp. From there the railroad

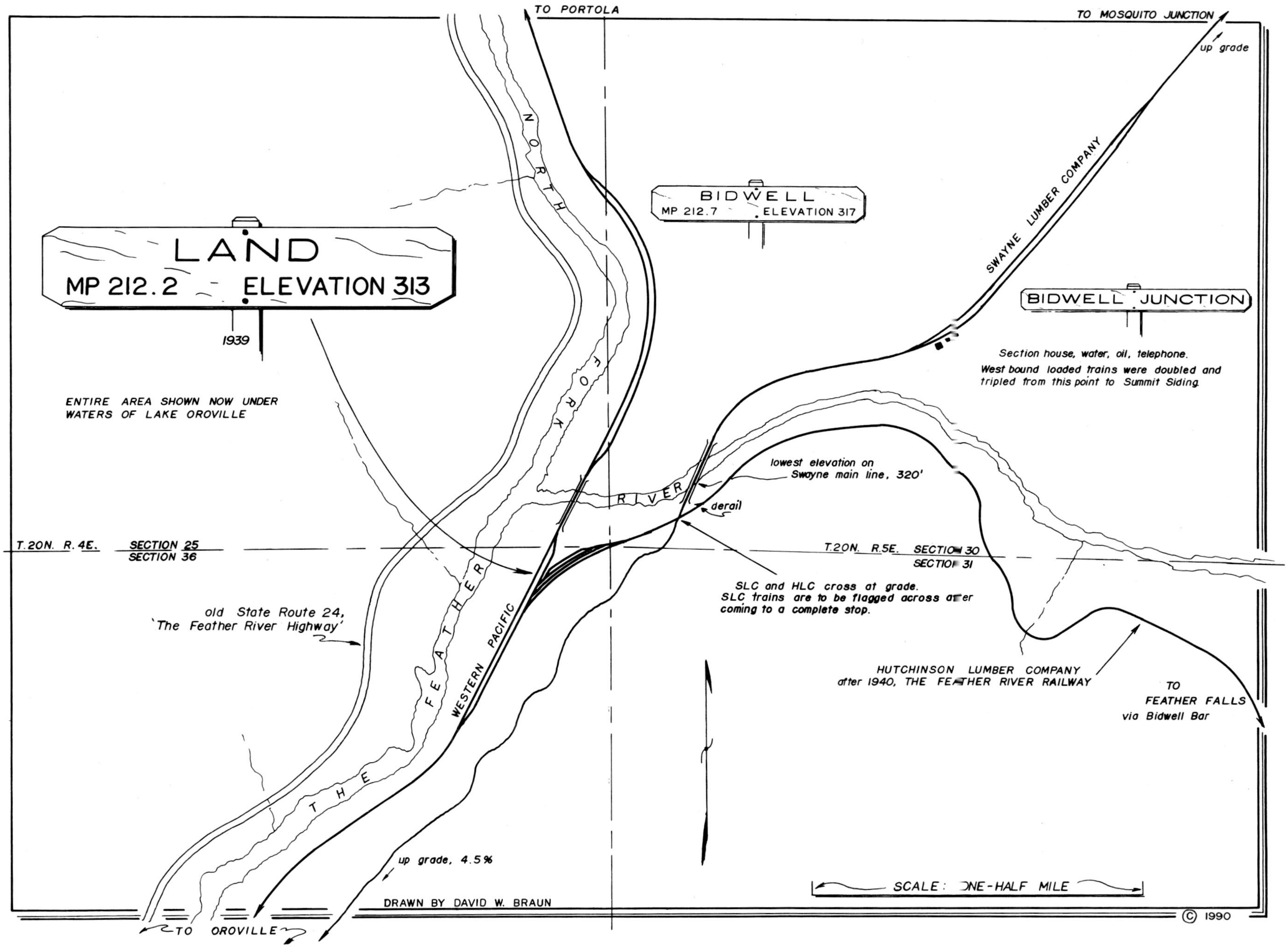

LAND
MP 212.2 ELEVATION 313
1939
ENTIRE AREA SHOWN NOW UNDER
WATERS OF LAKE OROVILLE
TO PORTOLA
TO MOSQUITO JUNCTION
up grade
BIDWELL
MP 212.7 ELEVATION 317
SWAYNE LUMBER COMPANY
BIDWELL JUNCTION
Section house, water, oil, telephone.
West bound loaded trains were doubled and
tripled from this point to Summit Siding.
NORTH FORK
THE FEATHER RIVER
WESTERN PACIFIC
lowest elevation on
Swayne main line, 320'
derail
T.20N. R.4E. SECTION 25
SECTION 36
T.20N. R.5E. SECTION 30
SECTION 31
SLC and HLC cross at grade.
SLC trains are to be flagged across after
coming to a complete stop.
old State Route 24,
'The Feather River Highway'
HUTCHINSON LUMBER COMPANY
after 1940, THE FEATHER RIVER RAILWAY
TO
FEATHER FALLS
via Bidwell Bar
up grade, 4.5%
SCALE: ONE-HALF MILE
DRAWN BY DAVID W. BRAUN
© 1990
TO OROVILLE

Here are several views of the spectacular Hartsmill trestle and ''Highway Bridge.'' At far left: No. 6 is stopped for water at the tank located on a little knoll between the two bridges. Below: A fireman's eye view of the Hartsmill trestle in 1928. The center photo shows Shay No. 1 backing downgrade on the Hartsmill trestle sometime in the 1930's. The water tank and ''Highway Bridge'' are visible in the background. The cabin to the left of the tank was occupied by a man paid to check the trestles for fire after the passing of trains. At left: A close-up of the ''Highway Bridge'' with the Oroville/Quincy highway running underneath. Above: Hartsmill trestle is being ''decorated'' by the bridge gang. Swayne trestles used three upright bends rather than four and train crews reported quite a few shaky moments while crossing. (Photos courtesy the California State Railroad Museum, E. E. Slaughter, Ed Thomas and J. E. Boynton collections)

branched three ways: to the west to Baker Creek; to the northwest to Coyote Gap, and to the east to Palmetto and Granite Basin. Wyes existed at Merrimac and at most other camps,[27] although employees recall they were seldom used.

Locomotives used in the 1920's operations were the two little Shays (3 and 4), the big three-truck engines (1, 2 and 6), Heisler 5 and the 0-6-0T "A." The "A," a wood-burning switcher, was used to dump log trains at the millpond. Big Shay 6, the "Cadillac" as the trainmen called it, was the Swayne pride and joy, arriving on the property in 1926. No. 6 was used initially on the "turn" to Mosquito Junction, exchanging MTy's for loads. Later, during the 1930's, the engine was run all the way to Granite Basin and the Little North Fork for the truck-to-train reloading operations.

In addition to the locomotives, the operation used one hundred log cars, three cabooses and several ancillary cars, the most important being the oil cars and the moving cars.

Moving cars were used to move the shovel, donkey engines, and later trucks and tractors, between the mill and the woods, as well as from camp to camp. The first moving car, the "Bull Car" as it was called, was built in 1925, using the trucks of wrecked log car No. 78 and steel I-beams. It was shortly joined by three ex-Southern Pacific forty-foot standard gauge flatcars, modified with narrow gauge trucks, three-quarter-size couplers and a second air line.

A story related by an ex-employee tells of the first time the larger cars were used. It was on a run to the Granite Basin to haul mining equipment on contract. The trip was made on a Sunday, ordinarily the crew's day off and the men were not very happy about having to work. With Shay 6, a caboose and the four cars heavily laden with equipment, the train left Oroville. It was very slow going, due in turn to the heavy load, the poor condition of the main line, and an unexpected problem — the width of the former standard gauge cars. The too-wide cars began scraping the sides of embankments and small trees close to the track. Finally, on a sharp curve, the sill of one of the cars became caught in a large pine tree, bringing the outfit to a sudden stop. Nothing more could be done until the tree was cut down and the stump dug up to free the train. After that trip the clearances at trackside were in effect "adjusted" by the scraping action of the big cars.

Log cars were used to carry more than just logs. Rather than adding a boxcar or a flatcar for incidental items that needed to go to camp, these items were thrown on the tops of empty skeletons for the return trip.

Ties were crisscrossed on the ends and bunks of the log cars, or laid lengthwise on the wooden main beams. Boxes, kegs and barrels were also hauled on the main beams, with only gravity and a few nailed cleats holding them in place. Fifty-five-gallon drums of kerosene and gasoline routinely made the trip between Oroville and camp in this fashion. One ex-employee remembers many times when they arrived in camp with fewer barrels than they had departing Oroville. He stated that "there must be a hundred barrels of fuel lying in the bottom of those canyons."[28] Another individual remembers a drum of gasoline "falling off" by being rolled off at French Creek by a train crewman. The barrel was hidden in the brush near a road crossing and was used to fuel his private automobile for Saturday night treks into Oroville.[29]

Train activities typically consisted of eight main line movements per day, six days a week. The woods switcher also worked six days per week and the crew went to work at 6 a.m. and tied up at 8 p.m., a long day, to say the least.

Train movements occurred in two divisions or "turns," the Oroville turn working Oroville to Mosquito Junction and the woods turn or "swing job" working from Mosquito Junction back to camp. There were two Oroville turns daily, both with eighteen to twenty empties and a caboose; the first leaving Oroville at 4:20 a.m., arriving Mosquito Junction at 7 a.m., and trading empties for loads with the first woods turn. The woods turn would leave camp at 5:30 a.m. with twenty loads, to arrive at the Junction by 7 a.m. Once loads were swapped for empties, both locomotives would return to their respective starting points.[30]

There were several interesting aspects of the operations of these daily turns. In the 1920's only the Oroville turn ran with a caboose, which stayed with the locomotive. The woods turn hauled only log cars, never utilizing a caboose. Despite the implication of the name, Mosquito Junction was a spur, not a siding. In order to exchange trains the engines would uncouple from the cars, leaving both trains on the main line. The woods turn locomotive, which had arrived first with its train of loads, would nose into the siding. The Oroville turn locomotive would run uphill, past the switch and couple against the loads. The woods turn engine would then run back down the siding out onto the main, and pull the empties and caboose back into the siding. Once the caboose was clear of the main, the switch was relined and the Oroville locomotive, with its new train of loads, would pull downhill and stop a car-length below the Mosquito Junction switch. The woods turn crew would then cut the caboose off from the empties, put their shoulders against the end sill and use gravity to roll the caboose down to the waiting loads. Once the caboose was coupled to the loads and an air test made, the Oroville turn would start back to Oroville. As soon as the Oroville turn left, the woods turn would back out of the

Swayne second No. 2 in service at Granite Basin. Notice close clearance to the trees. (J. E. Boynton collection)

A transition photo of No. 1 (originally ordered as No. 3) at the siding near bridge No. 46. The engineer was named Greenhill and the fireman was Charlie Anderson. (California State Railroad Museum)

No. 6 pauses at Stanwood Saddle in 1928. That's Charlie Anderson again in the doorway. No. 6's piston valve conversion has not yet taken place. (E. E. Slaughter)

A log train crossing Berry Creek at the site of present day Lake Madrone. (E. E. Slaughter)

Mosquito Junction siding and then reverse direction and start back to camp. The return trip to Oroville with the loads involved doubling or tripling the three miles of 4½-percent grade between Bidwell Junction and Summit Siding, plus a meet at "42" siding with the second Oroville turn, all of which kept the train crews and the dispatcher very busy.[31]

There was no wye at or near Mosquito Junction, so the Oroville turn wound up running the engine and caboose backwards to Oroville and the woods engine would have to run tank-first from camp. Since there was a considerable amount of standby time during the switching operation at the Junction, either one or both locomotives would take on Bunker C from the oil tank at the end of the siding. Operations were changed in the 1930's and cabooses ran all the way to the woods. The spur at Mosquito was lengthened into a regular logging spur about the same time, running approximately one and one-half miles westward.

Upon arrival at the mill in Oroville, the road engine would cut off from its train which was left standing on the main line. The mill switcher, usually the 0-6-0T "A," would then pull the train past the switch leading to the log dump, except for the caboose which was again left on the main line. "A" would then reverse direction and push the cars past the dump, one at a time. The caboose was later pulled into the yard and stored on the tail of the wye.

The second turn was treated in the same fashion and the tail track next to the log pond was long enough for two trains of twenty log cars each, where they were kept until the next morning's run. Occasionally cuts of log cars were also stored on the long spur next to the car shops. The cabooses were usually left on the tail of the wye where they were the last car coupled when the eastbound MTy trains were made up the following day. The cabooses were never turned and essentially spent their working life pointed eastbound. The Swayne mill yard trackage was minimal and did not allow for complex switching moves or large numbers of stored cars. The Southern Pacific dual-gauge trackage existed primarily for the box factory and lumber sheds, and the Sacramento Northern switched cars of mill ends and slabs.

Each main line train ran with a crew of five: an engineer, a fireman, two brakemen and a conductor. Main line train movements were dispatched from the Oroville office, using standard Form 19 train orders and a dispatcher's sheet. The company had a bare-wire telephone line that ran from the mill to camp and conductors had "hot stick" phone taps, called "caboose phones," that allowed them to call the dispatcher or camp. All of the telephones were of the hand-crank variety and the ones located along the tracks were housed in boxes. On one occasion, a trainman opened a seldom used box only to find a hornets' nest. He was stung a number of times before fleeing to the safety of the caboose.

On all main line train movements, the two brakemen "decorated the loads" by riding on top of the log loads. They spaced themselves six cars apart, when running with a caboose, or midway in the train and on the last car, if there was no caboose. The brakeman employed a unique application of the passenger train conductor's valve. Each brakemen had a "monkey tail," "tail hose" or "Joe Dwyer," as the safety devices were called. These devices consisted of a hose, connected into the automatic air line between cars by means of a pipe tee and gladhands, with an emergency valve on the other end and a hook to spike into the top log. This system allowed each brakeman, from his perch atop the loaded cars, to make a brake application as soon as a problem developed. Besides the engineer and the conductor, the safety factor was doubled, with four people having the option of stopping the train in an emergency.

Woods trains lacking a caboose placed a "monkey tail" on the rear car's gladhand and one mid-train. Often the woods trains were only nine or ten cars in length due to track conditions and gradients.

The practice of riding loaded log cars was not to be taken lightly, as Swayne did not secure log loads with chains. The logs remained in place by their own weight, resting against each other with adjustable cheese blocks on each log bunk. Cars were loaded to five thousand board feet each. It was the job of the head loader to carefully select each log for placement in a secure position. It was sarcastically spoken around camp that if a loader could not teeter a load with his shoulder, it was not properly loaded.[32]

That logs did roll off on occasion was supported by the fact that each Fall, a special clean-up train, equipped with the derrick, was employed to pick up fallen logs lying near the tracks. Often these fallen logs were the result of deliberate train crew actions. When a car would derail, crews would remove the log load by dragging the train to the next embankment and "jill poking" the load off by jamming a tie between the logs and the embankment and rolling the train ahead. This action reduced the weight so that the car could be rerailed more easily. Sometimes the entire car was "jill poked" off the track and left for the clean-up crew.[33] The practice of "jill poking" gave the clean-up crew more work time each Fall and it has been suggested there was some collusion between trainmen and the clean-up crew to increase the number of logs and cars to be picked up.

In retrospect, the "monkey tail" devices seemed an unusual system to ensure prompt response to problems and derailments. The employment of these unique devices was no doubt an effort to solve Swayne's continual

The Oroville mill, in operation with a full millpond, circa 1925. Hog feed on the right goes to the slash pile where waste is burned behind a concrete firewall. (J. E. Boynton collection) At right: A map of the Swayne mill area in 1926. (Sanborn Mapping & Geographic Information Service)

problems with extreme grades, fifty-degree curves, lost loads, derailments and rock slides.

An interesting aside to the practice of trainmen "decorating the loads" was the effects of inclement weather. Rubberized rain slickers were a very necessary wardrobe item. All of the three-truck Shays, as well as two-truck No. 3, had wooden lockers built either over or next to their rear sand tanks. These lockers were used to store the crew's raingear, as well as spare coupler pins, knuckles and air hoses.

Another aside is brakeman Harry Dennison's story relating to riding a log load through Merrimac. Another employee, Tom Blair, was hiding behind the Merrimac water tank waiting for Dennison. When the train rolled by Blair pulled the water valve rope and "damn near drowned" the brakeman.[34] After that incident Dennison was a lot more circumspect around locomotive water tanks.

Dennison also remembers coming through Merrimac one night and feeling a "bumpity bump." Realizing that the car he was riding had derailed, he "pulled the plug" (opened the emergency valve) and jumped. It was so dark he didn't see he was jumping into an embankment. Dennison ended up spread-eagled on the cut with a mouth full of dirt.[35]

Train crew workdays ranged from eight to thirteen or fourteen hours. All main-line movements were handled by one of the three-truck Shays. Two-truck Shays 3 and 4 and the Heisler 5 worked as woods switchers. Woods switchers stayed in the camps all season, with spot maintenance work being performed there. Boiler washouts on the woods engines were accomplished using a steam line from a donkey engine. Occasionally a mechanic would have to come out from Oroville with parts or welding equipment in one of the track cars if a locomotive developed a problem. On at least two occasions, locomotives with serious problems were "worked back" light to Oroville — the Heisler with broken gear teeth and one of the big three-truck Shays with a broken cylinder head. In the later Thirties, when the logging trucks began taking over, the woods switcher jobs were largely abolished.

The second set of train movements each day saw a second Oroville and woods turn again heading for a rendezvous at Mosquito Junction. The second Oroville turn would leave Oroville at 9 a.m., meeting the first Oroville turn as it returned westbound with its loads, and passing at "42" siding. The empty eastbound train arrived at "42" before the westbound train. The M1y would go into the siding and call the dispatcher at Oroville to let him know everything was all right. Obviously the eastbound train would not proceed from "42" unless he met the westbound turn or was given permission by the dispatcher.

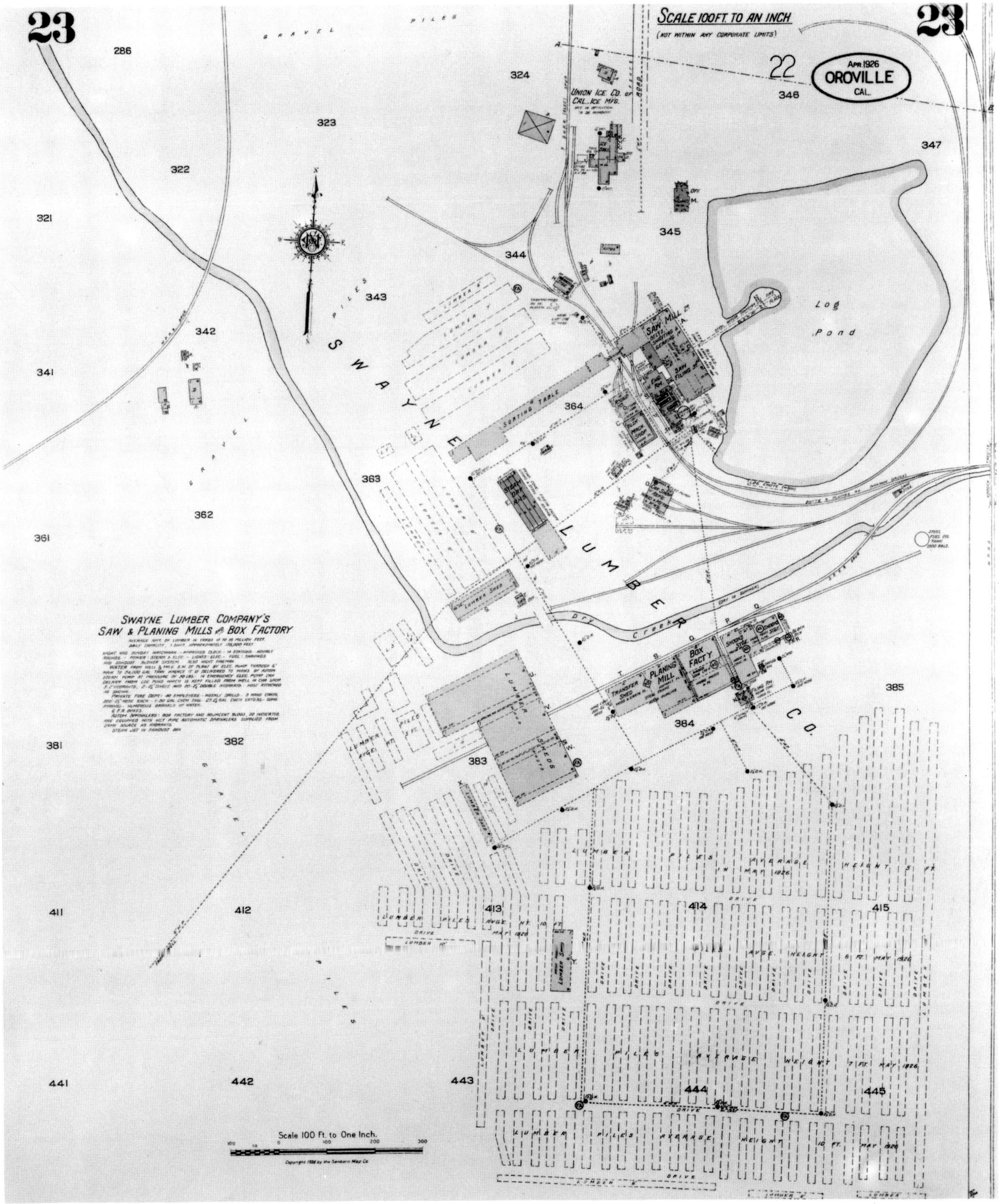
23
SCALE 100 FT. TO AN INCH
(NOT WITHIN ANY CORPORATE LIMITS)
23
APR 1926
OROVILLE
CAL.
22
286
324
323
322
321
346
347
345
344
343
342
341
364
363
362
361
Union Ice Co. of Cal. Ice Mfg.
Saw Mill
Log Pond
SWAYNE LUMBER CO.
Sorting Table
Dry Creek
Lumber Shed
Transfer Shed
Planing Mill
Box Fact'y
SWAYNE LUMBER COMPANY'S
SAW & PLANING MILLS AND BOX FACTORY
385
384
383
382
381
411
412
413
414
415
441
442
443
444
445
Scale 100 Ft. to One Inch
100
0
100
200
300

The second woods turn would be the same crew that brought the first woods turn, leaving camp at 10:30 a.m. and arriving at Mosquito Junction at noontime. Rather than swapping loads for empties like the first turn, the second turns would run by each other, pausing only to "gravity" the caboose out of the siding onto the rear of the Oroville-bound train. Engine crews would stay with their locomotives but train crews would trade trains.[36]

This system resulted in the cycling of an engine crew and a three-truck Shay between Oroville and camp daily. While this schedule meant a ten-hour day for the woods turn's engine crew, it also allowed for thirteen hours' rest the previous day and only one day away from home. The woods turn or "swing job" crews stayed in their own cabins at Merrimac Camp.

This system also ensured that the large Shays were cycled back to Oroville for inspection and maintenance at least every two days. The system was similar to the scheduling of operations on the Diamond & Caldor Railway running out of Diamond Springs.[37] Despite the extreme grades in and out of the Consumnes River canyon, no doubling was used on the D&C system, meaning a "turn" was actually two trains of ten cars each. And, unlike Swayne, the road engines were wyed at the junction where loads and empties were exchanged.

Like the Swayne operation, the Diamond & Caldor woods engine was not turned and one of the engines on the second turn was exchanged for the woods engine, thus cycling an engine back to the mill. The D&C also had the advantage of owning four three-truck Shays, which would keep one in standby status most of the time. The Swayne engines — all three of them — were used every day throughout the season and any breakdown was a true crisis.

Swayne locomotive engineers were Charles Anderson, John Bateman, Delmus Blair, Art Greenhill, Jack Crow, Bill Heath, Bill Otis Sr. and Al Johnson. Firemen were Blacky Strang, Harold Johnson, Virgil Edwards and "Butch" Slaughter. Conductors were Sam Crowley, Tom Fulk, Bill Cory and Henry Fulk. Brakemen were Harry Dennison, John Dresser, Tom Johnson, Frank Cory, George Rogers and Chester "Chub" Moore.[38] A quick survey of last names reveals the existence of nepotism in train-crew hiring.

Logging railroading was a dangerous occupation and the inevitable tragic accidents occurred. In 1919 a flatcar loaded with hay fell from Trestle No. 32 between Roger's and Black's Crossings, crushing brakeman John Boyd Dresser beneath the hay and killing him.

Sometimes death occurred because of habit or inattention. Conductor Tom Fulk, remembered as a "nervous, jumpy man" met death one day in June of 1928 precisely because of habit. Fulk didn't like riding in the caboose and had become accustomed, once the train left Oroville, to walk the twenty empty skeleton log cars forward to the locomotive to sit with the engine crew. Walking a thirty-inch-wide pathway is difficult at best, but compounded by the swaying of the cars and the slack action of the air brakes, it didn't take much to cause him to slip. Fulk fell from the train just upgrade of Bidwell Junction while checking groceries on an empty skeleton log car. The wheel flanges cut his body to pieces before the train could be stopped.[39]

The 1920's were the heyday of Swayne and its operations. Even as the 1929 logging season closed, no one could foresee the vast changes coming just within a year or two — changes that would affect not only Swayne, but the entire country as well. The logging industry in some ways was a microcosm of what was happening to industry and technology at large, and it would take a tough approach for a logger to survive the dark and forbidding 1930's.

Shay No. 3 is clearing snow near Palmetto, prior to the start of another logging "show" in the spring of 1935. Here, in several views, No. 3's snowplow crew takes time out to clown for the photographer. (J. E. Boynton collection)

Shay No. 1 on a caboose hop at Spooner Ravine, 1936. (J. E. Boynton collection) At right: A formal portrait of Lloyd Swayne, president of Swayne Lumber Company, 1925-1940. (Courtesy the estate of Lloyd Swayne)

CHAPTER FOUR

The Depression and Final Years for Swayne

The 1930's represented the final decade of Swayne's operations. It was a particularly challenging time for management, with the financial trauma of the Great Depression, the resulting New Deal reforms, rapid, far-reaching technological changes and Forest Service ecological requirements, all destined to forever change the techniques of the lumber industry.

The Depression hit the hardest between 1930 and 1933. Its effects were intensified in the western lumber industry by chronic overproduction, a situation that had been noted with alarm as early as January 1929.[1] Swayne necessarily cut production to sixteen million board feet, laid off a number of workers and hoped for better times. It was 1935 when business finally picked up again. The mere fact that Swayne Lumber Company weathered the Depression was probably due to the company's ties with Swayne and Hoyt, Inc. Lumber was shipped across the ocean to ports in other countries, allowing Swayne to fare better financially than most mills during these economically disastrous times.[2]

In the final years Swayne's management team consisted of Lloyd Swayne as president, Charles Schleef as

Swayne employed several different types of tractors with CARCO (PC&F) arches. Below is a Model 75 "Caterpillar" and at right, an Allis-Chalmers product. Swayne also used several different makes of logging trucks like this 1937 GMC tandem drive with Fabco dual axles. (Photos: Paul Beckstrom, Mallory Hope Ferrell collection and Courtesy the estate of Lloyd Swayne)

manager and Glen Bump as railroad general manager and logging superintendent. Lloyd Swayne became president in 1925 following the sudden attack of appendicitis and death of brother Warren H. Swayne.[3] Lloyd inherited the entire enterprise in August 1936 when his father Robert H. Swayne passed away.[4] Charles Schleef also took over in 1925 as manager. He began with Swayne in 1918, coming from the Meadow Valley Lumber Company at Grays Flat, Plumas County,[5] also the site of the northern terminus of the Spanish Peak Lumber Company's aerial lumber tramway. Schleef was destined to remain with Swayne to the very end, becoming the last employee on the payroll in 1940.

Glen Bump replaced railroad general manager and logging superintendent Pat Lyons in 1934.[6] Bump served in this capacity until the close of operations.

President Franklin Roosevelt's New Deal brought several changes to the Swayne workforce. Along with other western lumber operations, Swayne adopted a common industrial code under the provisions of the National Recovery Act. The code limited work hours to 48 per week, or six days per week, eight hours a day. Wages were also raised to $.375 an hour, with time and one-half for overtime. This rate replaced a pre-1933 scale of $.275 an hour.[7]

The most significant changes in Swayne's operations, however, were not the result of the Depression, but rather of new technology. World War One tank and diesel engine advances resulted in the introduction of both the Caterpillar tractor and the logging truck. The tractor, in combination with the development of the logging arch, spelled the demise of steam donkey engines and high lead cable logging by the mid-1930's. This change was also precipitated by U.S. Forest Service policy requiring Caterpillar tractors to be utilized wherever possible to lessen ecological damage. High lead logging was exceptionally damaging to smaller trees and young growth, and tractor logging represented a distinctly less destructive method of yarding and skidding.[8]

Swayne tried tractors as early as 1929. They also

Swayne utilized both Ross and Gerlinger lumber carriers to transport cut lumber in the yards and sheds. A garage was built to park the equipment when not in use. One driver gave his mount a customized look with a grill from a 1935 Ford truck. The Ross carriers were raised and lowered mechanically with chains while the Gerlinger's employed hydraulics. (Photos Courtesy the estate of Lloyd Swayne)

experimented with a gasoline-powered yarder that same year.[9] In 1931 the company increased its tractor equipment and added a bulldozer for road building.[10]

The June 1933 edition of *The Timberman* reported that Swayne was using Allis-Chalmers tractors.[11] Arches towed behind the tractors were used to lift one end of the logs above the ground to facilitate skidding. Swayne used CARCO arches, built by Pacific Car and Foundry in Renton, Washington.[12]

The mid-1930's also saw the introduction of Ross Carriers, or "carryalls," to the mill yard. These interesting gasoline-powered contrivances were utilized to move units of lumber from the green chain to the drying yards.

The logging truck was added in the 1938 operation out of the Little North Fork Camp. The truck, as in other logging shows through the country, was a final reason for terminating new railroad construction. The combination of tractors with arches, logging trucks and new logging road construction enabled logs to be brought down to railroad reload sites from locations previously thought to be inaccessible.

Areas of significant logging activity during the early 1930's were as follows:

—Gold Lake/Haphazard Creek area. A spur line was constructed to the west from Gold Lake Camp to access early Truckee Lumber holdings.[13]

—Upper Granite Basin/Frazier Creek area. This was also logged in the early Thirties, utilizing donkeys and high lead rigging. A spar tree was located near the Robinson Mine on Frazier Creek.[14]

—Peavine/French Creek area, accessed by a line built northwest from Stanwood Saddle, 1930-31.[15]

—Four Mile Ridge, between Peavine and Ram Creeks, was the site of yet another short spur (three miles). This area was logged prior to 1935.

The camps located in each area of activity were named Gold Lake, Upper Granite Basin, French Creek Road and Four Mile Ridge Road, respectively.

Mid-1930's operations centered around Coyote Gap

Coyote Gap skidding operations during the 1935 logging season. Tractors are Allis-Chalmers-Monarch 75's with CARCO arches. Dozer sports a Mack-Woolridge blade. (Courtesy the estate of Lloyd Swayne)

Camp. Charles Hobart, a local Indian, was the camp boss.[16] During the 1934-36 seasons, the various drainages surrounding Coyote Gap were logged, as was the northeast side of Big Bar Mountain. Saturday night revelers would hike the four miles down Mill Creek to the town of Pulga on the Western Pacific Railroad for the entertainment offered there.[17] *The Timberman,* January 1935 issue, described the Coyote Gap operations as follows:

> "Swayne Lumber Co., Oroville, logged 18,500,000 feet [sic] of sugar and white pine in 1934. G. B. Bump is logging superintendent. Three Monarch tractors are in service. Average skidding distance is 1,500 feet. Logs are loaded with a three drum tight line system. About 50 miles of narrow gauge railroad trackage is in service. Eight and one-half miles of track were built last year. Six miles of railroad will be constructed this year, employing 40 and 50-pound rail. Motive power includes four Lima Shays and one Heisler locomotive, three 50-ton [sic] and two 35-ton. Only three locomotives are in service. Maximum grade with the load is 4½ percent. The cars are equipped with straight air and automatic air on all cars. Key men in the camp are Charles Hobart, foreman; B. E. King, hooktender; E. A. Byrne, head loader and hooktender; R. H. Hartnett, Delmas Blair and D. M. Keadell, locomotive engineers; Clyde Mastellar, master mechanic; Earl Omholt, tractor doctor."[18]

In 1936 Swayne contracted with a well-known gypo logger, Fred Jenni, to tractor-log to the east of Merrimac Camp. This district had both old-growth and second-growth timber, but the trees were widely spaced and together with the rough terrain made it uneconomical for Swayne to log with their own crews. Jenni had previously worked for Fruit Growers Supply at Hilt, California. Swayne laid a one-thousand-foot spur for loading and Jenni shipped out two to four loads a day. He supplied his own equipment — two Best "60" tractors with CARCO arches and a modified Fordson tractor with a two-drum winch rigged to an A-frame for loading. Ken Sullivan, who drove one of the Bests for Jenni, remembered the crew of eight men were paid a flat 75 cents an hour. The water tank at Merrimac leaked water to such an extent that an informal shower house was set up underneath to allow the men to get the worst of the red dirt removed after work. Sullivan also commented that, due to the extremely rough country, the tractors could not often negotiate the terrain and Jenni had to borrow a Swayne donkey engine. Jenni did not provide board for the married men, who obtained their groceries by giving the Swayne train crews money to shop at the Purity Store in Oroville. The groceries would come up in crates on the empty skeleton log cars the next day. The Jenni contract was for the 1936 season only. Jenni himself was later killed by his own Best tractor at Castella.[19]

On October 1, 1936 a fire occurred on the mill grounds in Oroville. The inferno consumed eight standard gauge boxcars, the drying sheds and several million feet of dry lumber. Most of the loss was uninsured and therefore was a considerable financial setback for the company.[20]

The last two years of operation — 1937-38 — represented a significant transfer of operations to a location some ten air miles and several drainages away from Coyote Gap. This new area was known as Deer Park or Little North Fork Camp.[21] The move coincided with Lloyd Swayne's inheritance of the entire company following his father's death. A new railroad line was constructed into this area from near Merrimac. It ran northeast from the main line, beginning three miles below Merrimac at Junction House and followed the north slope of the Little North Fork of the Feather River drainage, terminating at the crossing of the Little North Fork at Deer Park, a total of three miles. A section house was constructed between Spooner Ravine and Marble Creek.

Little North Fork Camp occupied the end of this branch line. Today a U.S. Forest Service campground by the same name identifies this location. Little North Fork was a truck/rail reload, the last of its kind for the Swayne Lumber Company. Truck roads were built to the northeast to the headwaters of Coldwater Creek and up Bear Gulch in a southeasterly direction. A combination of tractors with arches and trucks were used to bring logs to the reload at Little North Fork. A gasoline-powered P&H shovel loader was used to transfer logs to rail cars for the 47-mile haul to Oroville.

An article in the November 1938 issue of *The Timberman* documents the Little North Fork show:

> "The Feather River of California is not descriptive in a way, as it and the numberless sparkling tributaries mark the terminus of the widely flung plateaus and prairie-like high mountain meadows to the north. The swift waters of this river mark a cleavage in the terrain so vast the eye of the traveler turns away in unbelieving bewilderment. It is a Paul Bunyan land brought to as near a reality as human mind can understand but no word can describe. But it surely is a country of he men and he logging. Up in the sugar and ponderosa regions of the Little North Fork, is the logging camp of Swayne Lumber Co., with mill and offices at Oroville. In keeping with this land is the logging superintendent, Glen Bump.
>
> "'So many wheel men have come here looking for work,' Mr. Bump explained, 'and after taking one look where the tractors and arches must work, they turn and head for other parts, that we've been training our own drivers right on the job and getting good results and practically no accidents.'
>
> "The results here, when the load-out sheet of

The Fred Jenni "gypo" operation at Merrimac in 1936 is well illustrated here. Best 60 tractors and CARCO arches belonged to Jenni (shown standing next to one of the tractors), as did the gasoline Fordson two-drum loader mounted on the donkey sled with A-frame. Note riveted bunk log car No. 72 and the litter of empty gas drums. The wood-burning Swayne donkey engine was borrowed by Jenni to reach areas too steep for tractor operation. A "long" camp house shows in background. The one view at Oroville shows the Fordson A-frame loader being readied for transport by "lowbed" with movable rear-axle assembly and "rooster," up the Oro-Quincy highway to Merrimac. That's a standard gauge sugar beet car and Sacramento Northern trolley wire in the background. Sugar beet cars were used for transporting slabwood, a mill byproduct, to fuel dealers in the San Francisco Bay area. (Courtesy the estate of Lloyd Swayne)

Pawling & Harnischfeger gasoline-powered shovel loader at a tractor/truck reload. Loading was accomplished with cable and end slings. (Courtesy the estate of Lloyd Swayne)

Bill Turpin, landing scaler was examined, showed better than 140,000 feet. Superintendent Bump stated that 125,000 was a fair average.

"Two L-O model Allis-Chalmers with Carco arches bring in this scale, Chad Vermillion and 'Wheelin' Jones, Drivers. Leonard Conger is skinner on another L-O Allis-Chalmers. Floyd Burns is tractor mechanic. T. R. Clark, on the business end of a big P&H shovel loader, swings the 32-foot and lesser length logs onto the trucks and trailers. O. Hendrix is head loader, Humboldt Shorty second loader. Five company trucks are on this seven-mile haul, GMC tandem drive with Fabco dual-axles, four on regular runs. Drivers are Ples Jones, Jim Smalley, James Dixon and Dan McGrew. Harry Horn is mechanic.

"Three tandem drive White trucks are also on this haul, in charge of Lloyd Carpenter. Dual-axle Isaacson trailers are used. Logs are tight-lined off the trucks and onto the car with spreader bar and straps; Gunnar Anderson head loader. Making a study of logging and a graduate of Oregon State College at Corvallis, L. J. Kendall, second loader, stated that there was one angle of logging he had never observed in print and one seldom spoken of. It is the ability of the loggers to perform their work all day through with hardly an order from the hook tender. Each and every one senses the next move, or he doesn't remain a logger.

"One worker here who has watched them come and go for nearly 30 years is Charley Anderson, grade boss. He, with Logging Engineer F. A. 'Silver' Compton and Superintendent Bump, figure out the grades for trucks and trains in this craggy and shaggy mountain wilderness. Important units in this work are two Allis-Chalmers tractors with Carco and Mack-Woolridge blades. One also pulls a Jumbo ripper, piloted by Art Hendrix, the other in charge of Charles Gibson, a finisher and dozerman of first rank. J. B. Purcell is filer and is one of the school who believes in very little swage to the raker. At the same time he defies anyone to show where he swages off a raker point. And he also requests the buckers not to keep the saws out more than two days, which they could easily do. And he doesn't want them brought in until the points are worn down enough so that constant jointing will not be necessary. He believes that it is much more to the company interest to wear the points off in the log than in the filing shack. Clifford Kernohan is bull buck, Lewis Prideaux in charge of the office."[22]

Another area of final operations was conducted off Shute Mountain, due east of the main line between Mountain House and Brush Creek. Tractors were utilized during the 1937-38 logging seasons. Logs were hauled to Sly Saddle, or Hurley's Saddle, and reloaded on log cars.[23] The reload method is unknown.

In the Spring of 1938 Manager Schleef attempted a shot in the arm for Swayne's declining box shook business. He worked with local Oroville businessmen to promote "Box Week," the purpose of which was to promote the extensive use of wooden boxes instead of less-durable cardboard ones.[24] Ads were taken out in the local newspaper touting the slogan "Keep the Whistles Blowing," referring to the whistles at the mill. Each issue of the paper during "Box Week" carried articles about the mill, its employees and its products.[25]

The box factory was Schleef's favorite of the Swayne operations. It employed forty employees at full production and Schleef's policy was to offer summertime employment to local high school students.[26] During its years of operation, box shook was produced for: two-section citrus crates, lettuce crates, vegetable crates,

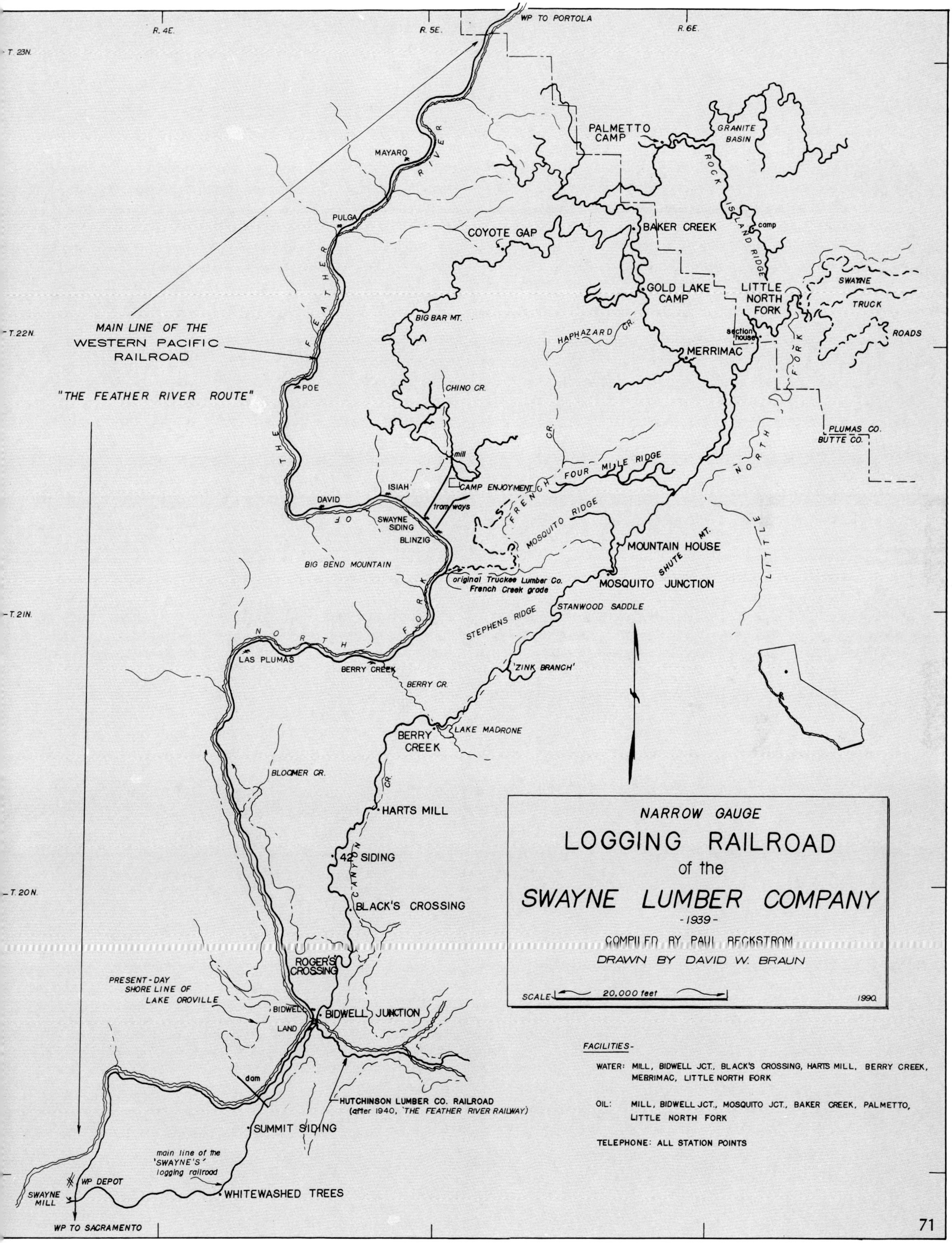

NARROW GAUGE
LOGGING RAILROAD
of the
SWAYNE LUMBER COMPANY
-1939-
COMPILED BY PAUL BECKSTROM
DRAWN BY DAVID W. BRAUN
SCALE 20,000 feet
1990
R. 4E.
R. 5E.
R. 6E.
T. 23N.
T. 22N.
T. 21N.
T. 20N.
WP TO PORTOLA
WP TO SACRAMENTO
MAIN LINE OF THE
WESTERN PACIFIC
RAILROAD
"THE FEATHER RIVER ROUTE"
THE FEATHER RIVER
NORTH FORK OF
MAYARO
PULGA
POE
DAVID
ISIAH
SWAYNE SIDING
BLINZIG
BIG BEND MOUNTAIN
LAS PLUMAS
BERRY CREEK
BERRY CR.
PALMETTO CAMP
GRANITE BASIN
ROCK ISLAND RIDGE
camp
BAKER CREEK
COYOTE GAP
GOLD LAKE CAMP
LITTLE NORTH FORK
section house
SWAYNE TRUCK ROADS
BIG BAR MT.
HAPHAZARD CR.
MERRIMAC
CHINO CR.
mill
CAMP ENJOYMENT
tramways
FRENCH CR.
FRENCH FOUR MILE RIDGE
MOSQUITO RIDGE
original Truckee Lumber Co. French Creek grade
MOUNTAIN HOUSE
SHUTE MT.
MOSQUITO JUNCTION
STANWOOD SADDLE
STEPHENS RIDGE
'ZINK BRANCH'
NORTH FORK
LITTLE NORTH
PLUMAS CO.
BUTTE CO.
LAKE MADRONE
BERRY CREEK
BLOOMER CR.
HARTS MILL
CANYON CR.
42 SIDING
BLACK'S CROSSING
ROGER'S CROSSING
PRESENT-DAY SHORE LINE OF LAKE OROVILLE
BIDWELL LAND
BIDWELL JUNCTION
dam
HUTCHINSON LUMBER CO. RAILROAD
(after 1940, 'THE FEATHER RIVER RAILWAY')
SUMMIT SIDING
main line of the 'SWAYNE'S' logging railroad
WP DEPOT
SWAYNE MILL
WHITEWASHED TREES
FACILITIES-
WATER: MILL, BIDWELL JCT., BLACK'S CROSSING, HARTS MILL, BERRY CREEK, MERRIMAC, LITTLE NORTH FORK
OIL: MILL, BIDWELL JCT., MOSQUITO JCT., BAKER CREEK, PALMETTO, LITTLE NORTH FORK
TELEPHONE: ALL STATION POINTS

L.A. Lugs, cherry boxes, Ghianda Ranch fig boxes, cheese boxes, Mt. Ida Oil boxes, Carnation, Alpine, Morning and Borden canned milk boxes, Hearst Ranch Fruits for Salads, Rural Gold Apricots, Ehmann Products, Apron Strings Fruit for Salads, and five-gallon can kerosene boxes.[27]

The 1938 logging season began in April with the employment of two hundred men, both in the woods and in the Oroville mill. The workers' time, however, was cut to seven hours a day. The box factory was operated only when it had orders. Operations continued through to the Fall. Winter saw the property shut down. In the Spring of 1939, instead of resuming work, Lloyd Swayne announced closure of the company for good.

Several factors contributed to the demise of Swayne's operations in the Merrimac Forest. The primary reason was a disappointing contract, let in 1938, for federal timber in the Granite Basin. The timber was well within the reach of Swayne's tractors to be hauled back to the Little North Fork reload; however, it was ridden with "shake," a series of vertical cracks following along the growth rings throughout the tree, making it unsuitable for lumber.[28]

Since the California lumber industry was in a market-demand mode during 1938 and 1939, economics played little part in the decision to terminate operations.[29] Swayne was stuck with a contract yielding unsuitable timber that had to be hauled some 47 miles to the mill in Oroville.

Swayne had also run out of privately owned timber and faced the prospect of buying additional land or contracting further with the United States Forest Service for federal trees. Forest Service contracts brought with them stricter logging regulations, resulting in higher costs per thousand board feet.

Finally, in 1938, the Southern Pacific debt that Swayne had assumed when purchasing the Truckee Lumber holdings in 1917 still amounted to approximately $200,000. This money could not be repaid so, for the second time in 27 years, the Southern Pacific reclaimed the property. The Swayne Lumber Company was finally released from its debt in November 1939 as the old Butte & Plumas Railway began to be taken up.[30]

Charles Schleef remained with the company until 1940 when SP's contractor, Hyman-Michaels, had completed tearing up the line and dismantling the mill.[31] The mill machinery was subsequently moved to the Winton Lumber Company mill in Martell, California.[32] Swayne employees scattered throughout the West Coast region, some going to lumber outfits like West Side, Michigan-California and California Door. Others were hired by the Western Pacific or Southern Pacific railroads.

In the end, Swayne Lumber Company neither went bankrupt nor experienced serious labor problems. The Swayne family simply terminated operations, closed the books and left town. With them went thirty years of narrow gauge mountain railroading; the machine gun-like exhausts of Shays echoing in steep wooded canyons; remote camps and colorful people, and a purpose and clamor never to be seen in the Merrimac area again.

The Merrimac Forest is now still. Second growth trees cover the scars left by high lead logging. A few camp houses remain, scattered here and there, serving as cabins and outbuildings. Wide, well-graded dirt roads have been built on top of the old roadbed and wood trestles have slowly turned to forest debris.

Shute Mountain show, 1937-38 logging season. At least two 1935 Ford trucks were utilized in conjunction with Cat 60 tractors, one of which was equipped with a two-drum winch. Reloading was accomplished with a wooden-boomed, gasoline-powered shovel loader. (Courtesy the estate of Lloyd Swayne)

The Swayne box factory and planing mill, with several adjoining sheds, was almost three hundred feet in length. Located to the south of the sawmill, the complex was fronted on its north side by the Southern Pacific spur. The two larger views opposite show the buildings from each end with completed boxes stacked for shipment. The bottom views form an almost continuous picture (looking west and from left to right) of the physical plant between the box factory and mill. The overhead pipe was used to send sawdust from the box factory back to the fuel bin at the corner of the mill. The machine shop and the open clerestory of the engine house can be seen in the last view. At left, rear views (south facing) of the box factory and planing mill are quite rare. The top photo shows the shook storage area with the concrete saw filing shack at right. A Ross carrier trundles next to the tramway which for some reason did not extend much further south to the dry yards. The lower photograph is the rear of the transfer shed next to the planing mill with the SP spur on the other side. (Courtesy the estate of Lloyd Swayne)

The box shook business was labor and machine intensive. These interior views, taken by Lloyd Swayne, show the various planers, edgers, bandsaws and other equipment used by his box factory. (Courtesy the estate of Lloyd Swayne)

The Swayne Lumber Company office was originally constructed in 1909 by the Truckee Lumber Company. It was built on a small mound of dredger tailings, placing it slightly higher than some of the other structures on the property. It was two-story with full basement and a large walk-in safe. The earlier view (at right) dates from the 1920's while the others are from the 1930's. The identities of the men are uncertain. The building itself was razed in 1989. (Courtesy the estate of Lloyd Swayne)

In the 1930's the dry yards were located west of the sorting table and south of the box factory. The lumber was hand-stacked in piles ranging from five to ten feet in height. The dry yards had no tramway and movement was accomplished by lumber carriers and, in some cases, horse and wagon. (Courtesy the estate of Lloyd Swayne)

SP MW70 17

Cleanup following the big fire of 1936 involved an SP "Big Hook." Remains of the boxcars attest to the intensity of the fire. Outside-braced boxcars of that era had wood sides, accounting for the grim skeletal remains. The sheds covered 20,000 square feet of floor area, limiting the fire fighting efforts to containment only. After the fire, the lumber and transfer sheds had to be rebuilt from the ground up. Here, several views trace progress from framing to finished buildings. The brand-new 1936 Pontiac next to the N&W boxcar helps to properly date the scene. (Courtesy the estate of Lloyd Swayne)

Little North Fork (Deer Park) extension, Spring 1937, with snow still lingering on the mountain ridges. By the 1930's railroad right-of-way construction had come of age, using jack hammers, tractors with blades and tracked shovels. Large view looks in an easterly direction up Little North Fork drainage toward camp. The portable air compressor powers the hammers and one of the Allis-Chalmers tractors prepares to push rubble down the mountainside. (Courtesy the estate of Lloyd Swayne)

GENERAL MOTORS TRUCK

Here are several 1937 GMC trucks used in the 1938 season. Above: While only a year old, this truck is already showing the rigors of logging service . . . missing bumper, multiple dents and crunched running board. The truck is spotted at the brow log in preparation for rolling logs onto the landing. Note unloading cables running over the brow log, under the load, and appearing again above logs at right. Cables will be pulled taught, causing the logs to roll to the left. Above left: a GMC truck at the reload. Operable windshields made cabs slightly more tenable by lessening dust, cooling somewhat and relieving the ever-present gas fumes. Lower left: At Bear Gulch or possibly Coldwater Creek, a truck with a gasoline-powered two-drum loader operating tight line with spreader. Also shown is a GMC truck with a P&H shovel loader. Note the missing door on the driver's side. (Courtesy the estate of Lloyd Swayne)

Swayne's water wagon was a 1931 Diamond T, shown here at an unknown truck/rail reload. This truck was used for a variety of tasks, including fire fighting, watering the donkeys, and keeping the roads wetted down. Note the spotlight for night operations and the 1½-inch fire hose with nozzle. Right above: A 1936 White at Little North Fork reload. Drivers doors were often removed for better air circulation and less impediment to the driver. Large skids on the earthen mound at the top of the photo belong to the donkey loader. Right below: Tight-lining a load of logs on one of the GMC logging trucks at an unknown reload location. (Courtesy the estate of Lloyd Swayne)

The unloading, storage and movement of logs to the mill's jack slip is shown in this series of photographs. The logs were tight-lined off the log cars by a spreader bar and two cables. The "brow log" was actually scrap railroad rails. Once in the millpond, the logs were moved either by men with pike poles or by a carriage with tongs suspended from a cable. Swayne did not have a pond boat. The cable ran from an A-frame next to the unloading frame, to yet another (steel) frame located a the end of the jack slip. Both A-frames had small two-drum engines powered by a steam line from the mill. Here, the venerable 0-6-0T "A" pushes the logcars into position. The logs are rolled off, then moved to the slip and finally conveyed into the mill after being washed. (Courtesy the estate of Lloyd Swayne)

The Swayne mill at Oroville at peak production in the 1930's. A full log pond tells of maximum output by the woods crews. The pump house in the concrete structure next to the jack slip provided water pressure for the log washer, a close-up of which is shown above. (Courtesy the estate of Lloyd Swayne)

The log pond was kept clean of bark and other debris by a homemade dredger using a conveyor belt. The muck was off-loaded into boxes on rafts, as seen in these views. This was done to prevent the filling in of the pond bottom by debris, eventually resulting in a pond too shallow to float logs. (Courtesy the estate of Lloyd Swayne)

Above: Sometimes the logs were washed by a worker with a hose, as shown here. The small shed in the foreground is the pump house and one of the last of the riveted bunk log cars has just been unloaded, leaving a small amount of slash on the beams. Right above: The sawdust bin was located in a reinforced concrete structure next to the engine and boiler room. Water for the boilers was located next to the bin in a 54,000-gallon wooden water tank. Right: Link-belting for mill machinery sits neatly rolled next to the engine house lead. (Courtesy the estate of Lloyd Swayne)

This view, looking northeast, is one of the few photographs showing the electric repair shed (at left). The machine shop and engine house with its open-air clerestory are immediately behind. In the foreground is a stake body from a truck and, at the extreme right, the carpenter shop. (Courtesy the estate of Lloyd Swayne)

Crawler-tracked, gasoline-powered shovel languishes under lumber shed. Builder's plate reads: "The Brown Hoist Machinery Co., Cleveland, Ohio." (Courtesy the estate of Lloyd Swayne)

One of the oil cars sits on the Sacramento Northern dual-gauge trackage next to the fuel bin and power house. In the background is one of the Shays under the overhead conveyor to the sorting table. At right: The yard office was covered with a specialty siding cut to resemble logs, a Swayne product. The office was located between the drying kilns and the lumber sheds. Left above: The Swayne engine house was used for repair work other than locomotives due to the availability of an overhead crane. Here a six-cylinder gasoline tractor motor and transmission is being overhauled next to Shay No. 2. Above right: Swayne locomotives were watered at Oroville by a two-thousand-gallon water tank near the engine house lead. The tank itself was a commercial type sold by Diamond Match for ranches and farms. Here, looking to the northwest, is the water tank, woods warehouse at the throat of the wye, the little dump cars used for ballast work, and the Sacramento Northern lead. The conveyor to the sorting tables is overhead. (Courtesy the estate of Lloyd Swayne)

The sorting table was located at right angles to the mill. Green cut lumber was delivered to the table by an overhead conveyor. The table was 260 feet long with sides left open to allow handling the lumber, but with a roof to give some protection from the hot summers so common in the Oroville area. The stacks of lumber will be taken to the dry yard by carrier or horse and wagon. At the far end of the sorting table, under the large roof, was the tramway and transfer table for lumber to be kiln-dried. The dry kiln was one hundred feet long and constructed of reinforced concrete. (Courtesy the estate of Lloyd Swayne)

Slash and other debris from the mill was burned in the open air behind a fire wall, the fire being fed by a conveyor. The wall was located on a spit of land that extended out into the millpond. The large view above looks north towards the olive plant and Table Mountain. The smaller pictures are looking southeast and the Swayne oil tank, holding fifteen hundred barrels of fuel oil, together with the Western Pacific bridge (under which the Swayne main line ran) can clearly be seen. (Courtesy the estate of Lloyd Swayne)

A westbound Western Pacific freight rumbles by the Swayne mill on a quiet Sunday afternoon. In the foreground is a cut of twenty empty log cars, ready to go out Monday morning. The derrick is seen on the next spur, with the boxcar parked by the car shops. A sugar beet car sits on the Sacramento Northern spur next to the sawdust bin.

At left: Pacific Gas & Electric transformers for service to the north side of the property were located next to the woods warehouse and the narrow gauge wye. Tank car No. 3 sits in the background amid stacks of cedar ties. A millyard "China cart" lumber conveyance is in the foreground. (Courtesy the estate of Lloyd Swayne)

After the lumber sheds were rebuilt in the fall of 1936, it was back to work loading lumber into boxcars from such varied roads as the Milwaukee and Southern. This inside view reveals no automatic sprinkler system. Fire protection was limited to hose standpipes and a manual fire alarm system. (Courtesy the estate of Lloyd Swayne)

The mill was powered by a Corliss steam engine with boilers fired by sawdust piped from the mill and box factory and stored in the concrete fuel bin. These interior views show the engine, generator and pyrometers. From this point, both electricity and steam were wired or piped to various areas and structures on the mill site. At left: The shotgun carriage in the mill was overhauled in the 1930's. Here, new hydraulic arms to turn the logs during sawing have just been unloaded and are sitting inside on their shipping pallets. (Courtesy the estate of Lloyd Swayne)

Above: A spar tree in the process of being rigged. A high-climber rides the cable to a new position. When completed, the spar will be utilized for high-lead yarding operations. (Courtesy the estate of Lloyd Swayne)
Above right: Raising a spar pole near Big Bar Lookout in 1937. This operation was near Coyote Gap Camp. (California State Railroad Museum)

A Swayne steam donkey on the "bull car" on Big Bar Mountain, 1937. (California State Railroad Museum)

Above: A log train in the woods at Fagan Saddle. Position of the caboose with its cupola forward corroborates ex-employees' statements that cabooses were not turned in the woods. Below: A log train doubling the five-percent grade on Bidwell Hill at Summit Siding in 1937. (California State Railroad Museum)

Here are several views of Shay No. 6, taken on May 30, 1938. Above: The engine is switching at Oroville before leaving town with a train of empties for the woods. At right: The rear brakeman (under the SP's wooden bridge in the background) is "rolling the train by." Second bridge back belongs to the Western Pacific. Switch and dual gauge was how the SP accessed the box factory, planing mill and lumber sheds. Note the boxes, drums and timbers on empties en route to camp. (Ted Wurm photo above; all others W. C. Whittaker)

6

6

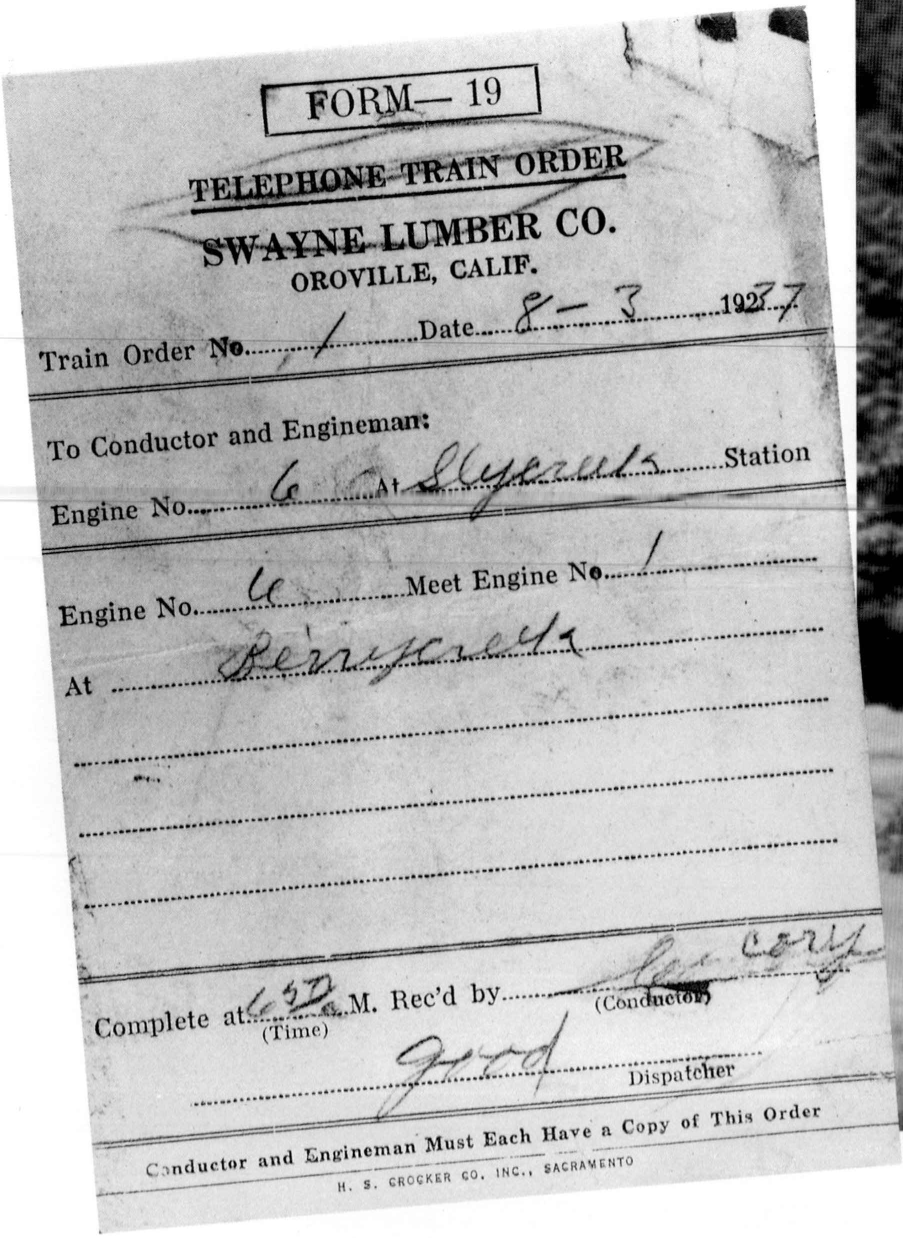

FORM— 19

TELEPHONE TRAIN ORDER

SWAYNE LUMBER CO.

OROVILLE, CALIF.

Train Order No. 1 Date 8-3 1937

To Conductor and Engineman:

Engine No. 6 At [illegible] Station

Engine No. 6 Meet Engine No. 1

At Berrycreek

Complete at 6:30 .M. Rec'd by [illegible] (Conductor)
(Time)

Good

Dispatcher

Conductor and Engineman Must Each Have a Copy of This Order

H. S. CROCKER CO., INC., SACRAMENTO

Above: Swayne order form 19, a telephone train order setting up a meet between engines No. 6 and No. 1 at Berry Creek. (J. E. Boynton collection) Above right: Shay No. 1 at Stanwood Saddle on September 8, 1939, while at right, No. 6 switches at Berry Creek. (California State Railroad Museum)

1

6

Scrapping operations for Swayne began on March 17, 1940. Here, Shay No. 3 is assembling the tear-up train. The fire fighting car is in front of the locomotive with small two-tank car ahead. This car was used to transport gasoline and diesel fuel to camps. At right: No. 3 is poling the fire fighting car after an unsuccessful drop left the car fouling the switch. The tracks had not been used for several months and had to be carefully tested before operations began. The next day, March 18th, the fire fighting car and tank car (No. 3) were destroyed in a runaway at Whitewashed Trees. (J. E. Boynton; California State Railroad Museum; Ted Wurm opposite)

3

SWAYNE LUMBER CO
3

The Southern Pacific did not find many customers for the former Swayne equipment, since the number of narrow gauge railroads had dwindled by 1940. The best pieces — the three-truck Shays and log cars — went to the last of California's narrow gauge logging outfits, West Side and Michigan-California. Other equipment, such as the 0-6-0T's, found no buyers. The last piece of equipment to leave Oroville was the little rotary snowplow. It was purchased by the Nevada County Narrow Gauge in October 1940. The photograph by Al Phelps shows the plow arriving and being unloaded at Colfax, California. Shays No. 2 and No. 6 were shipped to their new owners on standard gauge trucks in February and March 1940, while the No. 1 went to its new home, Michigan-California Lumber Company, on a standard gauge flatcar. In this photograph by Larry Harrison, the locomotive pauses in SP's Sacramento yards on February 18, 1940. (Photos at right: Al Phelps; California State Railroad Museum)

6
SWAYNE LUMBER COMPANY

The standard gauge Sacramento Northern had other customers to the north of the Swayne mill, including Union Ice, a gravel plant and an olive processing plant. Here, motor No. 405 switches Pacific Fruit Express reefers at the olive plant. The SN-WP interchange track is in the foreground and the lead to the Swayne mill, under trolley wire, is at extreme left. No. 405 was a homemade locomotive, built in 1914 by the predecessor Northern Electric at Chico. Sold in 1953 to Caldor Lumber Company, it provided trucks and other parts to an equally homemade diesel, built in the company shops at Diamond Springs. No. 405 was the only interurban freight locomotive to be on the property of two narrow gauge loggers in its lifetime — Swayne and the Diamond & Caldor. (Jeffrey Moreau collection)

Opposite page: Swayne's steel bridge over the Middle Fork of the Feather River, looking northwest. This structure replaced the original wooden span built in 1910. (J. E. Boynton collection) After abandonment, the SP's scrapping contractor took out the center span of the bridge. Here, in December 1942, we see the former Swayne bridge, while in the background the WP's *Scenic Limited* passes the station point at Bidwell, en route to Oroville. This photograph was taken by the California Highway Department which was surveying the old bridge as a possible single-lane throughway for traffic on the east side of the Feather River's North Fork. Despite some preliminary work, however, nothing further was done. This area is now under the waters of Lake Oroville. (John Snyder collection)

Little North Fork Camp in 1939. Brow log for crotch line truck/rail reload and one spar tree appear alongside the locomotive. Several "short" camp houses are in the background. (Paul Beckstrom collection)

CHAPTER FIVE

Camps and Camp Life

The logging camp was an integral part of Swayne Lumber Company's operations, from Stanwood in the beginning until the closing of Little North Fork Camp in 1939. Due to the rugged Sierra terrain and the lack of good roads, camps were a necessary means of housing workers within a half-mile's walking distance of the timber to be cut. Even when speeders or "crummies" were added in the late 1930's to transport the workers, the remoteness of the area dictated the continued need for the camps.

Camps housed anywhere from sixty to 150 men, depending on the number of "sides" or loading locations being utilized. While originally bachelor camps, in later years tent platforms and canvas shelters were provided by the company for workers with families, and camp population grew accordingly.

A typical listing of camp inhabitants during active timber-cutting periods is listed below. This crew would be necessary to support five donkey yarders.

Fallers	10
Buckers	30
Choker Setters	10
Choker Hole Diggers	5
Whistle Punks	5
Donkey Engineers	5
Donkey Firemen	5
Wood Choppers	10
Teamsters	5
Hook Tenders	5
Loaders	10
Head Riggers	5
Night Watchmen	5
Donkey Doctor	1
Pipe Man	1
Cable Splicers	5
Donkey Fireman for pumping donkey	1
Saw Filers	2
Total	120 men!

To this list must be added various support personnel who were necessary to make a camp function. The camp boss, cooks (head cook, second cook, and baker), dishwasher, waiters (called "flunkies"), bull cook and the timekeeper also lived in camp. There was also a blacksmith (in the 1920's) and a welder (in the 1930's). Railroad section gangs were also often located at camp. Finally, the woods switcher crew had a cabin and most camps also accommodated the overnight layover of the woods turn or "swing job" crews at one time or another.

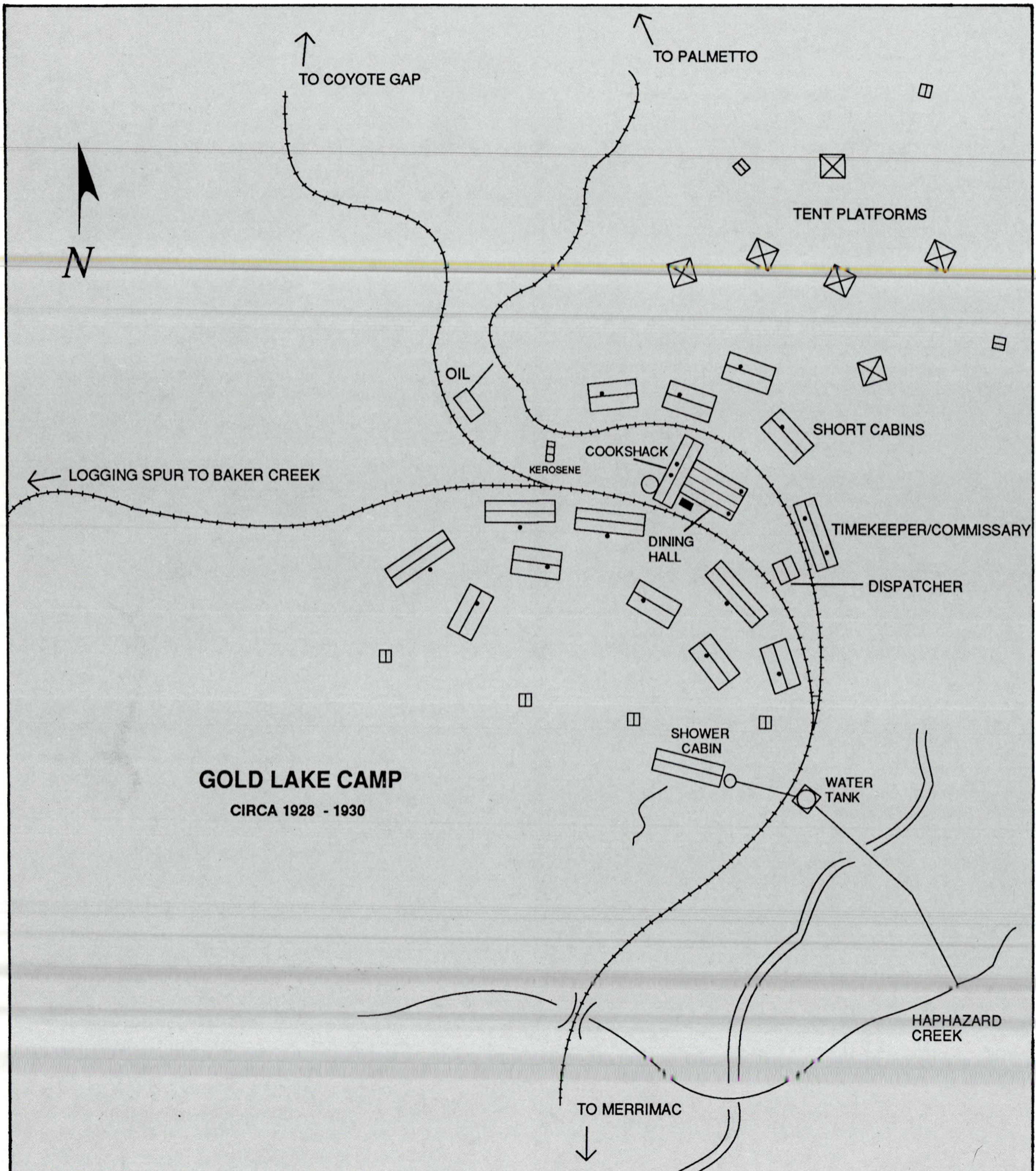

Map by Bob Mariano

Camp locations moved over the years as the timber cutting moved. Some camps remained in one location for several years and others for only one season. Camps were left in place each winter after being "winterized" or closed up by the last workers out. Walter Sturgeon of Oroville relates the story when he and three other men were staying at Gold Lake Camp the last night prior to closing up for winter. They were playing cards in one of the camp houses while a storm raged outside. Suddenly a large ponderosa pine crashed through the cabin, just missing the loggers and destroying the building.[2]

Camps were located in accordance with state public health regulations for logging camps. The following represents requirements typically in force after World War One:

1. Camps shall not be located at a distance less than two hundred feet from a river, brook or lake.
2. The source of water supply shall be a stated distance, generally one hundred to two hundred feet from and above, in elevation, any possible pollution.
3. Every structure serving as a habitation shall be furnished with ventilation flues and windows of sufficient number and dimensions to ventilate and light the interior properly. In some states, a minimum of six hundred or more cubic feet of air space must be provided for each occupant of sleeping quarters.
4. Floors must be elevated a minimum distance above the ground; side walls must be a minimum height; beds must be elevated above the floor a stated distance and must be separated at least two feet apart.
5. Adequate facilities must be provided for excluding flies from the kitchen and dining rooms.
6. There must be a system for proper disposal of garbage and refuse, such as burning or burying a minimum distance from the nearest camp building.
7. Latrines shall be located, constructed and cleaned in accordance with stated specifications.
8. Provision must be made for isolating men with contagious diseases.
9. Penalties are usually provided for infraction of all rules by a properly constituted state official.[3]

Sometimes water had to be pumped some distance to camps that could not be located near a spring or stream. One-cylinder gasoline pumps or old donkey engines were used for pumping the water through 1½-inch pipe.

Swayne Lumber Company had at least thirteen camps over the years: Construction Camp, French Creek Road, Merrimac, Four Mile Ridge Road, Gold Lake, Baker Creek, Sly Creek, Mosquito Creek, Palmetto, Rock Island Ridge, Upper Granite Basin, Coyote Gap and Little North Fork (Deer Park). Interviews with former employees indicate generally only one camp existed at any one time.

Track was laid through the new campsite and the buildings were moved over the railroad to the new location. Once the timber ran out, the camp houses would be loaded onto skeleton log cars (used to move long cabins), or flatcars (for short cabins), and hauled to the next site. In later years, trucks were used to move the camps.

The earliest camp houses were apparently built between 1917 and the early Twenties. The cabins, which were eight feet wide and 28 feet long (long cabins), could fit on the bunks of a skeleton log cars. Each cabin housed four men, with two bunks, bunkbed style, at each end. A table, benches and shelves were provided in the center section, along with a wood-burning stove for heat. Ex-employees referred to their "lockers" as being their suitcases, shoved into a corner or under a bunk. Two of these early cabins survive today in nearly original condition at a mine near the site of the Little North Fork Camp.

The newer camp houses (short cabins) were wider but shorter and offered eighteen square feet more floor space. This newer style also had a steeper roof pitch than did the original, which helped to shed the snow in winter. The cabins were all mounted on 16"×16" skids. Inside were four metal beds, one in each of the four corners, a woodstove against the rear wall and a table and benches in the center. Two washtubs sat on a bench outside the cabin for washing and shaving. Cabins were painted oxide red on the outside only.

Accommodations were often less than ideal. As one ex-employee train crewman relates: he arrived at Granite Basin Camp and discovered that the bedbugs were so bad he took his bedroll and slept on the tender deck of one of the Shays.[4]

There was one short cabin that was sheathed on the inside and kept freshly painted. It was reserved for visiting dignitaries, company officials, or Swayne family members. Apparently this cabin at one time was painted green, as it was referred to as the "green cabin."[5]

The typical Swayne camp consisted of a dozen or more cabins set up as terrain and stumps dictated. The dining units, cookshack and commissary were centrally located, near the tracks to expedite the unloading of supplies. A shower cabin was located on one end of camp. It was provided with an elevated water tank and a crude system of coiled tubing that allowed the water to be warmed by a wood fire. Shower water drained directly onto the ground beneath the cabin.

The woods switcher crew had their own cabin near the tracks, and yet another cabin was reserved for the layover woods turn or "swing job" train crew. The woods switcher tied up directly in front of the dining hall.

The timekeeper had a cabin near the cookshack. The loggers had to check in daily with the timekeeper in order to insure their timecards correctly reflected each day's work. The timekeeper also tended the commissary, where goods could be purchased by the workers. Items stocked included calked boots, tobacco, Levis, Frisco jeans, soap, toilet articles and bandaids.[6] The timekeeper also issued mattresses to each employee, who was required to furnish his own bedding in the form of a bedroll.

Workers paid the company ten cents per day for rent and one dollar per day for food. Swayne camps had the reputation of providing large quantities of quality food at each meal.[7] Breakfast consisted of hot cereal, pancakes, bacon, ham, either fried or boiled eggs, fried potatoes, hot biscuits and canned fruit. Lunches were packed for the loggers in boxes. The usual lunch was three sandwiches: cheese, ham, and jam, together with pieces of fruit. Dinner was roast beef, stew, macaroni or rice, potatoes, fruit, baked bread and dessert. Porkchops or steak was served two or three times a week, and fish was served on Fridays. Coffee was served, with condensed milk for cream, in thick crockery mugs without handles.[8]

The cookshack and dining hall was actually three skid cabins placed together. The dining hall was comprised of two long cabins, placed side by side. Each cabin lacked one wall, respectively, with only 4″×4″ posts holding up the roof, so when placed together, the dining area was as unobstructed as possible.

Meals were announced by the ringing of an iron triangle called the "gut gong."[9] Men sat on benches at wooden tables. Each man sat in the same place at each meal. Train crews sat at their own table near the front door. At Granite Basin, hogs were kept under the dining hall and were fed cookshack scraps. Train crewmen remember taking cold pancakes, rolling them up, and stuffing them down knot holes at breakfast to give to the hogs.[10]

The cookshack attached to the end of the dining units. It contained two cook stoves with right and left hand fireboxes placed together, so that one man could stand between them and stoke both stoves at the same time. In later years, the cookshack contained a walk-in coldbox kept cool with block ice brought up in a boxcar. During the Twenties, perishables were kept cool by digging a hole in the ground and lining it with block ice or by constructing crude evaporative coolers by draping burlap sacks over crates, placing vegetables inside and allowing a water pipe to drip onto the burlap and evaporate.[11]

A small elevated water tank was provided next to the cookshack for potable water. It was gravity fed, or pumped with a one-cylinder distillate pump.

Foodstuffs came up once a week from Oroville in a boxcar at the rear of the train. Coal for the cookstoves came piled on a flatcar. Coal was used in the cookshack because of its superior cooking qualities. Kerosene or "white gas" for camp lanterns came in 55-gallon drums placed on empty skeleton logcars.

In later years, a gasoline generator was provided for lighting the dining hall and cookshack. Loggers did not care for this convenience, however, as it was so loud "it would wake the dead."[12] The generator, therefore, was turned off about eight or nine o'clock each night.

The cookshack crew consisted of a head cook who did all the meat-cutting and cooking, a second cook who did the frying and a baker. The cooks and servers all wore flour-sack aprons. The "bull cook" took care of firewood for the camphouse stoves, coal for the cookshack stoves, kerosene for the camp lanterns, and also washed the wool blankets used on beds.

Cookshack crews had their share of interesting characters. One crew at the Palmetto Camp habitually got complaints that the men received no syrup with their canned fruit. It was later discovered that the cooks were siphoning off the juice to use in their liquor still, set up behind the camp.[13]

In another memorable incident, Lloyd Swayne was visiting camp. The cook did not know Swayne (and visa versa), nor had he been warned that the boss was coming to camp. When Swayne asked the cook what he was doing with such a dirty apron, the cook answered: "I'm diddling the dog; that's what I'm doing." Mr. Swayne did not like the cook's flippant reply and answered, "Do you know that I'm the dog you've been diddling? You are fired!"[14]

Since smoking in the woods was prohibited due to the fire danger, the loggers chewed tobacco during their workday. Smoking was allowed only in camp or in the cleared areas around the donkey settings. The donkey settings were specially prepared so that sparks from the donkeys' smokestacks would not start fires.

Drinking water was supplied in company-provided "Desert Brand" canvas water bags which were hung in the shade and shared among workers. One former employee indicated his preference to drink from the many springs in the woods. He justifid his choice by stating: "I'm telling you, it was something to get ahold of a bag that some big old Swede who had a mouthful of snooze had just taken a drink out of."[15]

Loggers engaged in horseplay and practical jokes, like those in every other line of endeavor. One former employee related a joke played by himself and one other worker on one of the railroad conductors. It was common practice to roll cuts of empty log cars into landing sidings, or onto the main, by gravity. This involved the timely throwing of switches. The two loggers, who had discovered a den of rattlesnakes and had killed a number of them, wrapped the largest dead snake around a strategic railroad switchstand handle. During the gravity switching operation, the conductor failed to operate the switch upon seeing the snake, and stopped, frozen with fear. The cut of cars rolled on by the switch, down to the end of a spur, and off the end of track into a canyon. The train crew spent two full days getting the derailed cars back out of the canyon and onto the track. Needless to say, no one ever owned up to the "joke."[16] This incident is believed to have involved log car 78, which was so badly wrecked that it was never rebuilt, the trucks and other parts being utilized in the construction of the Bull Car in 1925.

After dinner, in the evenings, the loggers had several hours of leisure time before bed. Showers were taken in the shower cabin, although some men only bathed on Saturday nights. Some men were so tired that they retired to bed immediately after dinner. Others would do wash, mend clothes, play cards, gamble or sit around and drink bootleg liquor and "tell big lies."[17] Camp kerosene lanterns generally gave off too poor a quality of light for much reading.

Saturday night, the end of the workweek, was a night of celebration, especially on payday, which was every other week. Some men rode available automobiles into Oroville or went down into the various Feather River canyon towns for entertainment.

Bootleggers and prostitutes often came up from Oroville to sell their wares. The prostitutes, usually four in number plus a madam, would set up tents within a half-mile of camp. Once several loggers' wives who lived in camp went down to the prostitutes' camp with a loaded shotgun and threatened to kill them all. The Saturday night revelry was cut short that night as the girls beat a hasty retreat back to Oroville. At least two paydays went by before they returned.[18]

Despite the loggers' proclivity for drinking, gambling, whoring and fighting, there was one vice that was simply not tolerated by anyone in the camps and that was thievery. Stealing from a fellow worker was the cardinal sin. No one locked cabins or suitcases when they left in the morning for work and all expected to find their possessions in the same place when they returned in the evening. Anyone caught stealing from the cabins, or even suspected of taking something for personal gain, could expect to be run out of camp at the very least. Many culprits were badly beaten on their way out. One ex-employee stated that in 1929 a rumor had it that one thief was murdered and the body stuffed down an abandoned mineshaft. For, as he put it, "the Oroville Sheriff was a long way off and the men didn't want an outsider to interfere in their business."[19]

Swayne "long" camp house at Coldwater Creek in 1983, still used by miners. (Paul Beckstrom collection)

Swayne Lumber Company Camp House (Original)

Scale: 3.5mm = 1 foot (87:1 HO scale)

Drawn by A. G. Wren.

Drawn by A. G. Wren.

Measured and photographed by Paul Beckstrom and A. G. Wren, June 1983.

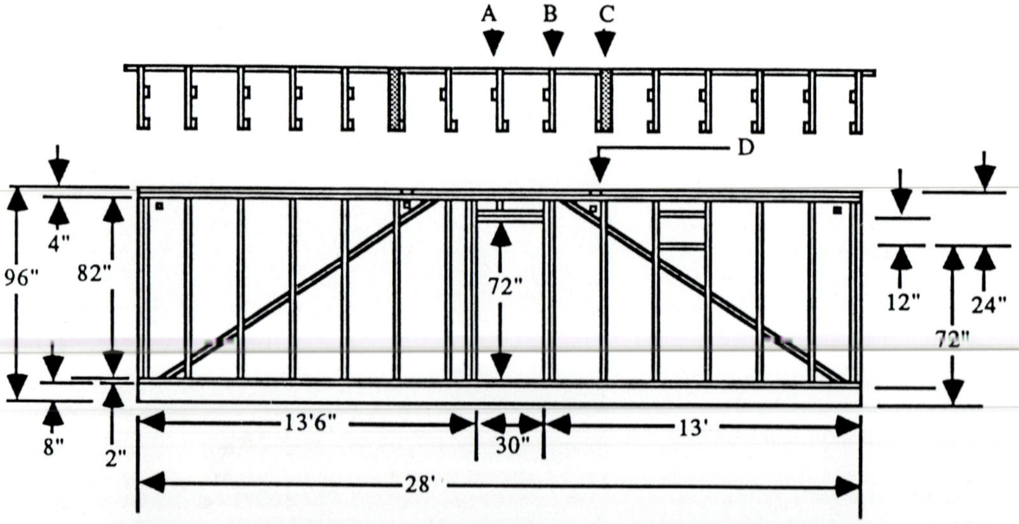

All framing 2"x4" on 2' centers. Roof trusses are 2x4 rafters with 1x4 bracing unless otherwise noted.

A,B,C and D crossections: see Figure 1.d.

Framing and siding details vary from car to car. Drawings are a composite of two existing cars.

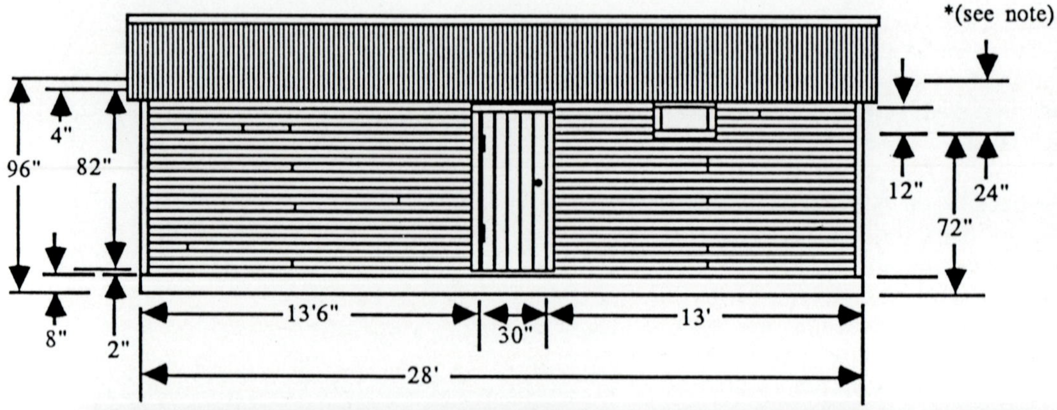

All trim is either 1x4 or 1x2 unless otherwise noted.

Door on this car only is 30" wide x 72" tall; window is 1'x1'8" and is not usually found on Swayne cabins.

Original roof was tar paper; current roof is corrugated metal. Roofing is laid vertically.

Side view, back, framing and siding

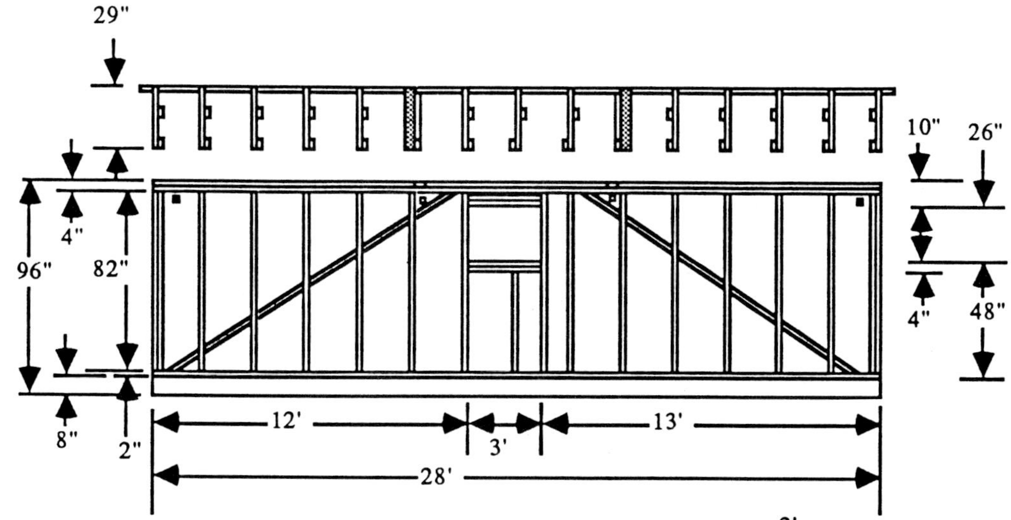

Window detail: see Figure 1.e.

Siding on original cabins was 1x3 or 1x5 drop siding

All dimensions true; rough-cut lumber used.

Cabins did not have rafter tails as of 1983. We do not know if the cars originally had them. Later cars do have normal rafter overhang.

1/2" dia. truss rods (4/cabin) were used for bracing.

2'x3' plate for chimney

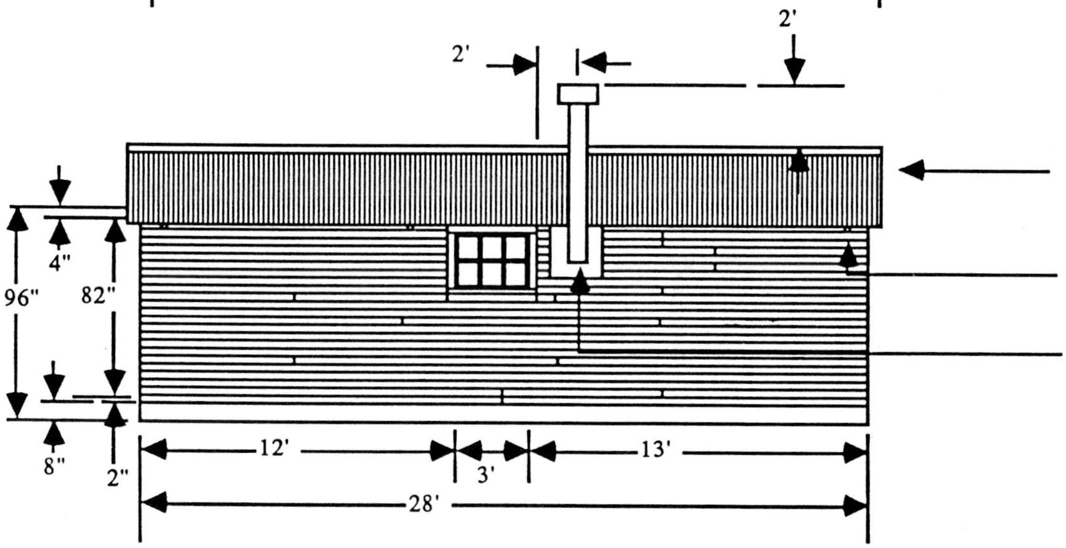

*note: Drawings show roofing overhang that was found on cars at date of measurement. We believe that these longer (older) cars were built without roof overhang. Thus roofing would end at rafters, 10" shorter than this drawing shows. Indicator line for top plate can be used as a reference for original roofing.

Opposite: A second "long" camp house. This one is constructed of 1×5 rather than 1×3 drop siding. Right: A close-up of the "long" camp house at Coldwater Creek. (Paul Beckstrom collection)

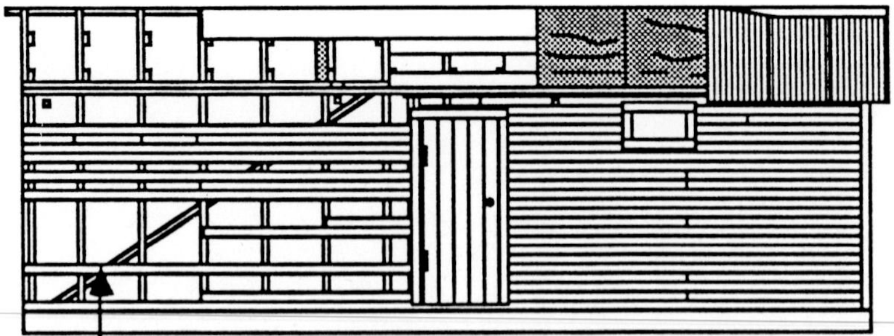

1x3 drop siding

Roofing is 1x3 to 1x12 random sheeting covered with tar paper. Current roof is corrugated metal with a smooth metal cap.

Roof truss detail:

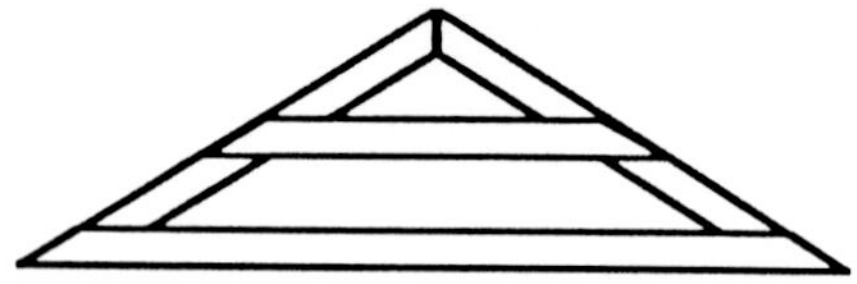

One of the existing cars has ends identical, while the other has them as mirror images (the window on both ends is skewed to the rear of the car, and they are directly opposite each other)

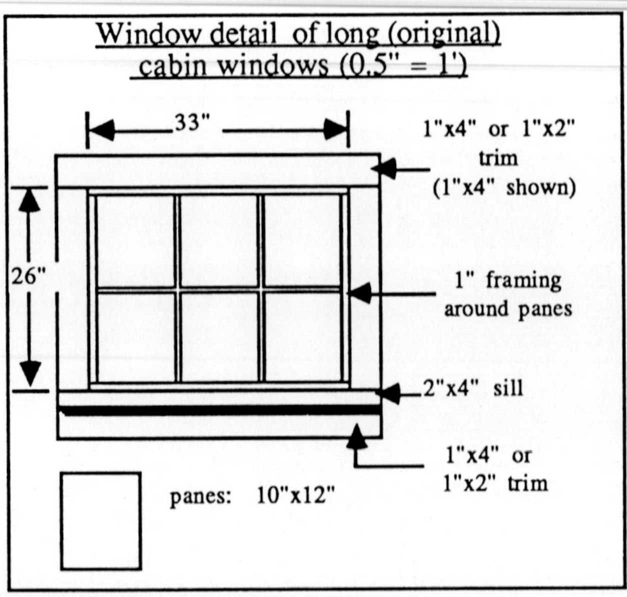

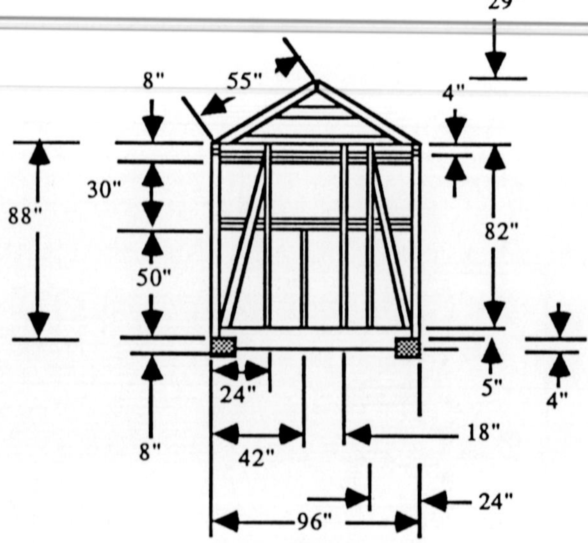

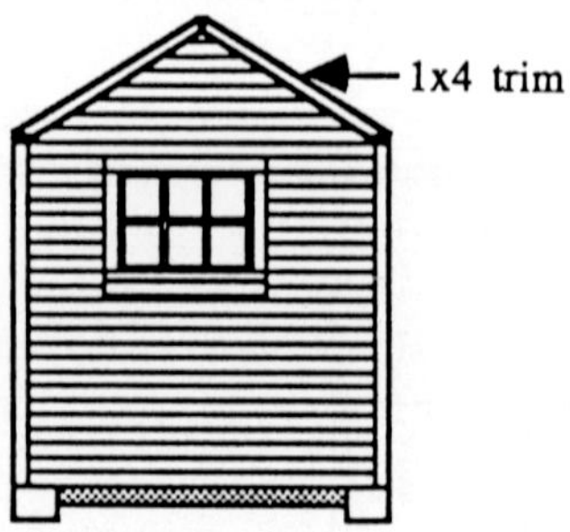

Rear view of the same cabin. These structures, with bunkbeds built into each end, housed four men. (Paul Beckstrom photo)

"Long" cabin at Buck's Lake. This cabin was once part of the cook-shack/dining hall. (Anthony Wren)

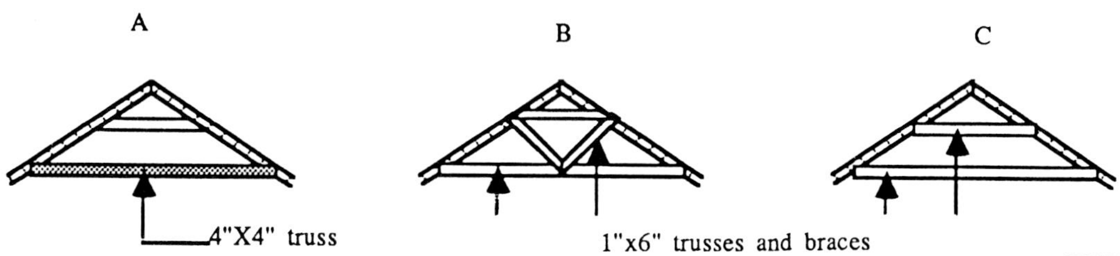

RAFTER DETAIL

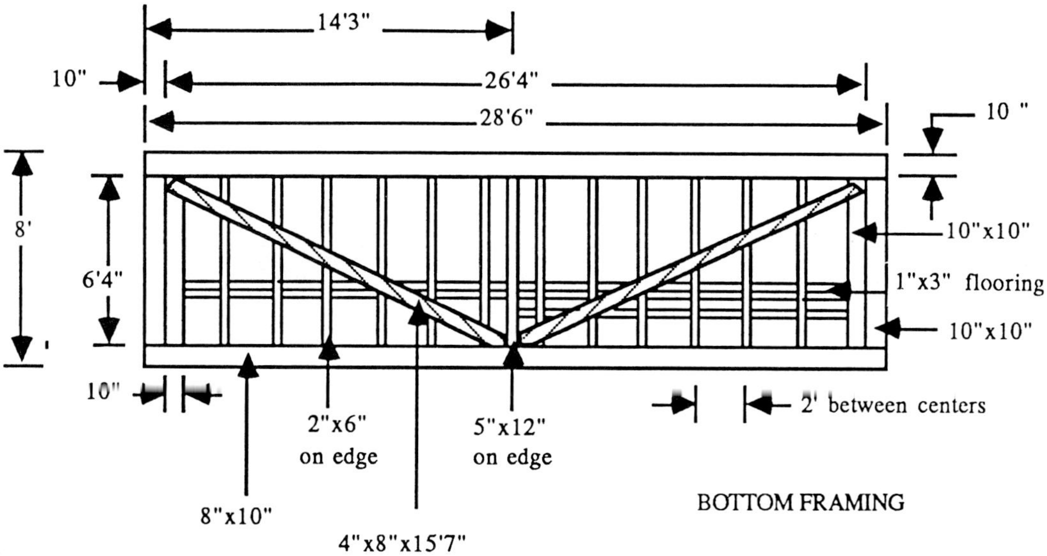

BOTTOM FRAMING

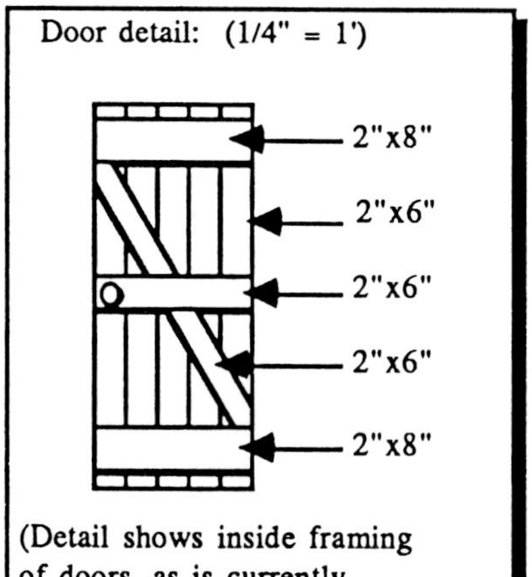

(Detail shows inside framing of doors, as is currently found on existing short cars.)

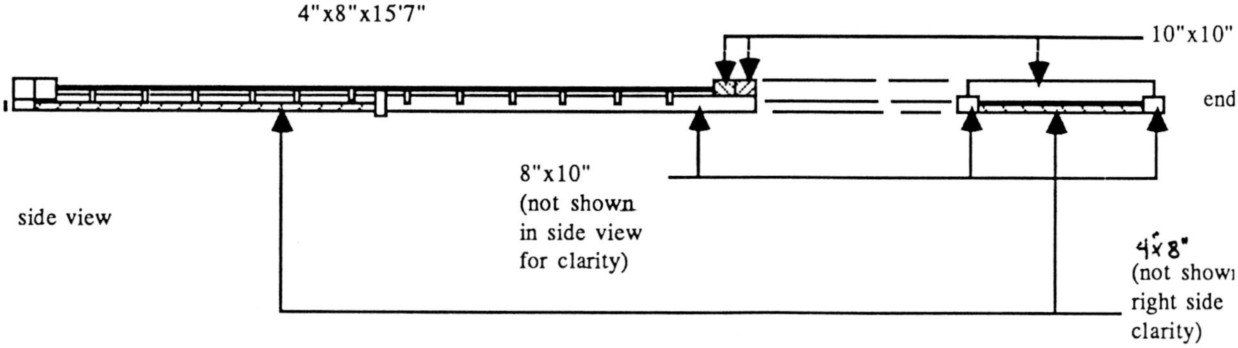

side view

Side view, external

* Rafters are 2"x6" on some cars, 2"x4" on others. This car is shown with 2"x6" rafters.
Modern versions do not have stoves or chimneys.
Variations occur, as indicated.

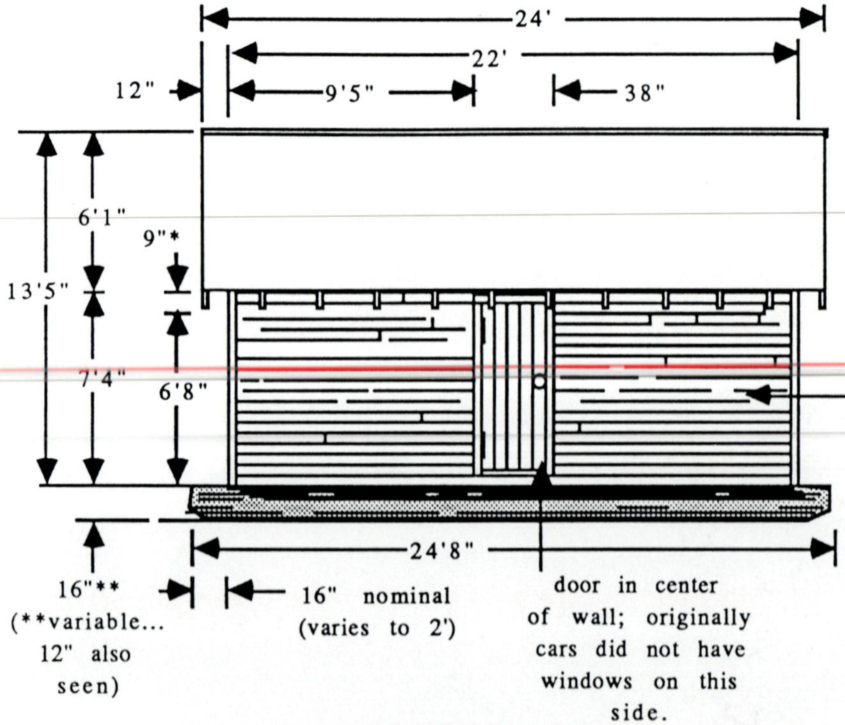

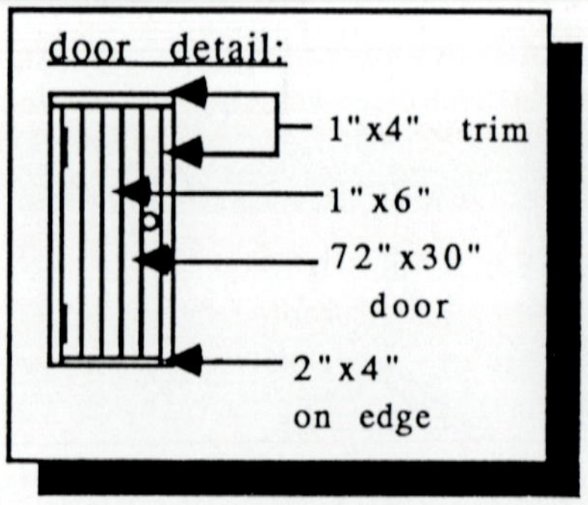

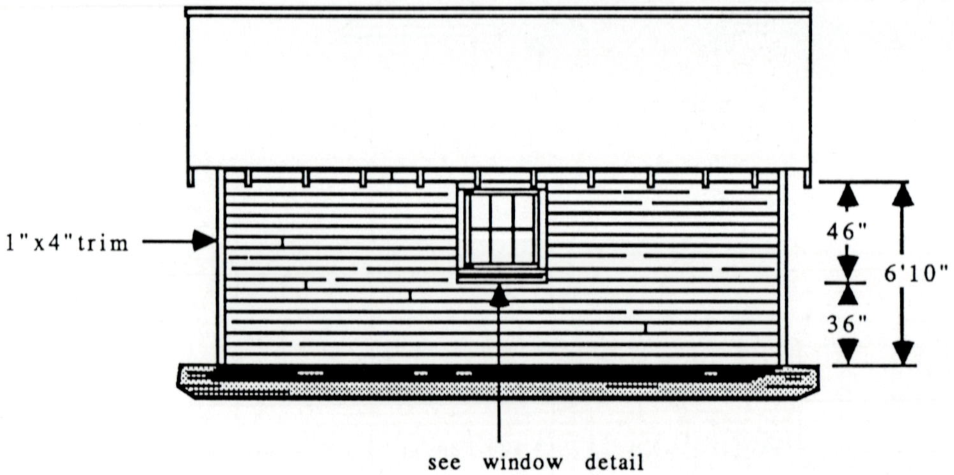

Skid detail (1" = 16")

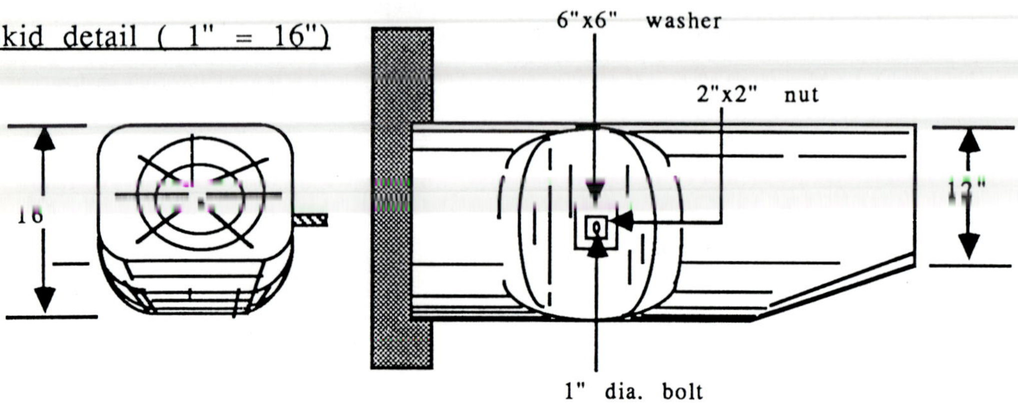

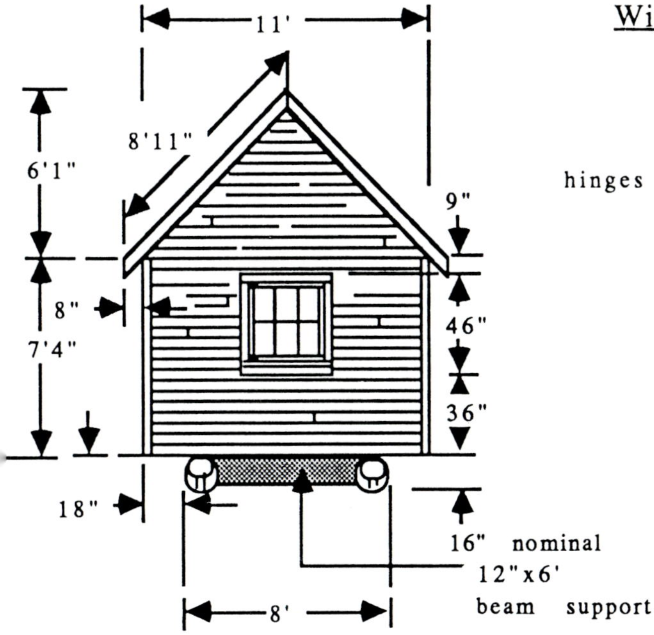

Edge trim is 1"x4"
Rafter and roof facia is 1"x4" or 1"x6"
Siding is 1"x5" V-rustic

Window detail (0.5" = 1') of short cabin windows

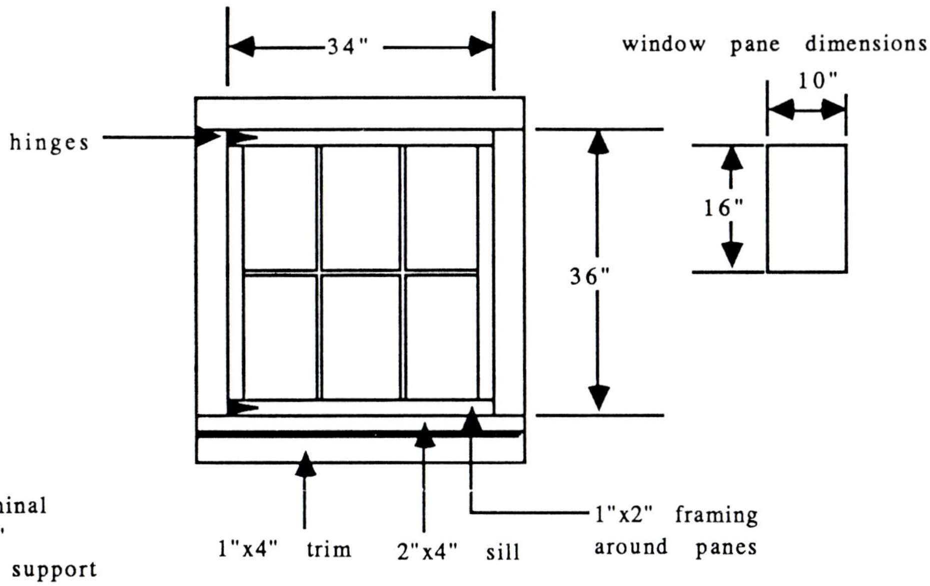

All dimensions true (rough-cut lumber used)
Roofing is tar paper (original) or corrugated metal (modern)

Swayne Lumber Company Short Camp House

Scale: 3.5mm = 1 foot (87:1 HO scale)

Drawn by A. G. Wren.

A "short" cabin with original skids still in place. At right: another example of a "short" cabin with one window added in the front. (Paul Beckstrom photos)

Swayne Lumber Company Shower Cabin

Scale: 3.5mm = 1 foot (87:1 HO scale)

Drawn by A. G. Wren.

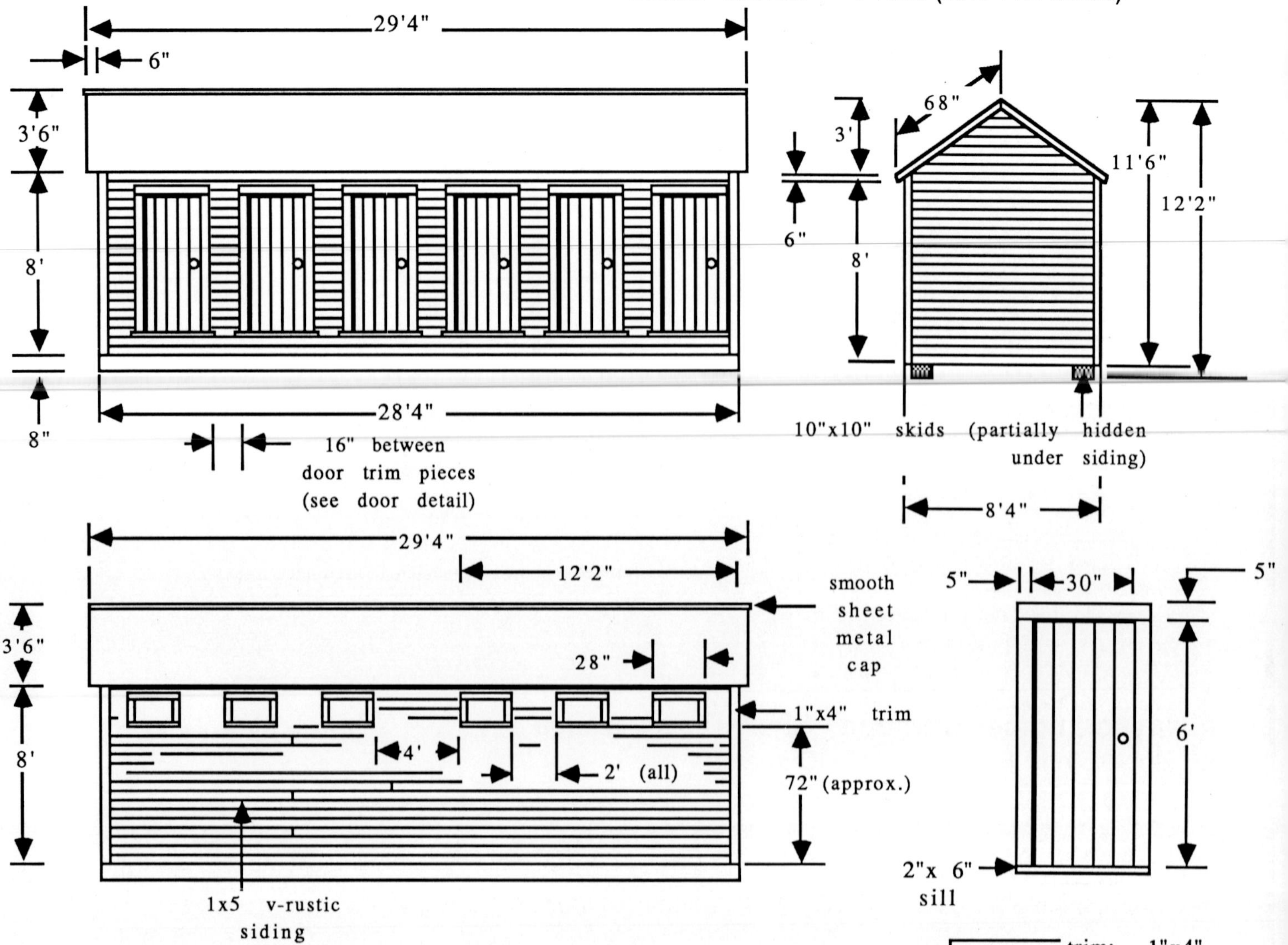

Window Detail:

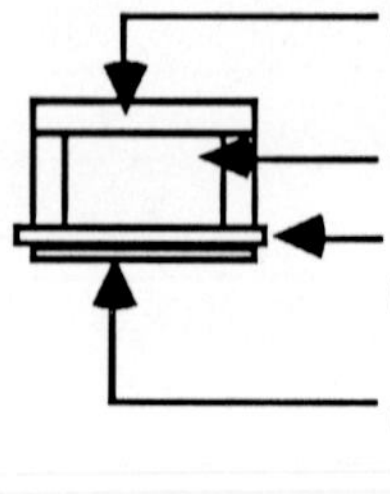

Edge trim is 1"x4" (ends have trim under eaves)

Rafter and roof facia is 1"x4"

Siding is 1"x5" V-rustic

All dimensions true (rough cut lumber used)

Roofing is tar paper (original) or corrugated metal (modern)

Remains of the shower cabin with original six doorways still evident, despite remodeling. (Paul Beckstrom)

The Remains . . .

In this photograph by veteran rail photographer Guy Dunscomb, Western Pacific No. 94 is at Land on October 9, 1954 during a railfan excursion. The WP main line is visible at left; loco is sitting on joint Feather River Ry./WPRR trackage. The concrete piers of Swayne Lumber's Middle Fork bridge appear just above the smoke stack. Swayne crossed Hutchinson Lumber Company track at grade just off to the right of the photograph.

For the adventurous railfan, remnants of Swayne's narrow gauge railroad can still be found throughout the Merrimac Forest; not so the Swayne office building. In this photograph, taken in 1988, the structure sits boarded up, on top of the hill overlooking the millpond. In 1989 the building was demolished and the hill leveled. The concrete fire wall, once separating the burning slash pile from the mill, is the only remaining vestige of the Swayne mill. The millpond depression is directly behind the fire wall and further back is the Union Pacific (ex-WP) main line. (Both photos: Paul Beckstrom) On the opposite page, above, is the roadbed remains of the North Fork line just east of Junction House. Today one has to be careful since this is a favorite trail for bears. At far right, above, is Feather River Little North Fork country. This photo, taken in 1989, attests to the rugged and remote character of the area logged by Swayne. The picture below, taken from the roadbed of Bidwell Hill just east of Summit Siding, graphically shows the level of the railroad in comparison with the top of Oroville Dam, seen looming in the background. At right: trestle remains with bents still upright can be found near Canyon Creek. The rusted steel box in the next photo actually housed a flange oiler, a device placed on opposite sides at each end of a major trestle. Both oil and water were used at various times in an effort to reduce friction. (All David Braun photos)

ROLLING STOCK ROSTERS
The Swayne Lumber Company and Associated Firms

TRUCKEE, CALIFORNIA AREA / The Truckee Lumber Company, 1867-1909

No specific records known. Photos show a mixture of flatcars, disconnected trucks and small "Russell-type" skeleton log cars. With the arrival of the Lake Tahoe Railway & Transportation Company in 1900, TLC operations in the Ward Creek and Squaw Valley areas utilized some of the LTRy&TCo. 24-foot "taper sill" flatcars which had previously seen service over at Glenbrook and Bijou. These flatcars remained the property of the LTRy&TCo. and were furnished to TLC on a lease or per-diem basis, with the Lake Tahoe Railway providing train crews and locomotives to haul the logs over their main line at 28 cents per thousand feet. *(See main text for freight rate discussion.)*

When TLC operations were shifted to Oroville, some of the equipment was taken to the new site. As far as is known, none of the LTRy&TCo. flats were purchased or moved, as Railway Commission reports show the class remaining intact through 1919 at Lake Tahoe. It is probable that some equipment formerly belonging to the Pacific Lumber & Wood Company was moved to Oroville with the two 0-6-0T's. It is not known if anything from the Cantara operation was sent to Oroville.

OROVILLE, CALIFORNIA AREA

Note: It is important to keep in mind that the Truckee Lumber Company and the Swayne Lumber Company were independent firms, operating in two different locations in the general Oroville area prior to 1917.

The Truckee Lumber Company (Sept. 1909 to Sept. 1911) — Operating as The Butte and Plumas Railway

An unknown amount of equipment was brought up from Truckee at the start of operations. This equipment was supplemented as follows:

In December 1909 the South Pacific Coast placed many of its flats, boxcars and some passenger cars on what they called "The Laid Aside List." Equipment on this list was available for sale. Some of this equipment was original SPC; while some was originally from the Oregonian (Oregon & California) or the Pacific Improvement Company.

Between September 1910 and February 1911, five lots of cars were sold from the "Laid Aside List." Three of these lots were sold to the Butte & Plumas Railway at Oroville:

Lot "A" — November 1910; 30 flatcars and 3 boxcars.
Lot "B" — January 1911; 10 flatcars.
Lot "C" — February 1911; 30 flatcars, 1 boxcar and 1 passenger car.

Lot "C" was noted as being sold to "Butte & Plumas Ry. Co. (Truckee Lumber Company)."

The boxcars were SPC No. 190 and No. 196 (28-foot, 15-ton) and No. 358 (28-foot, 10-ton) for Lot "A," and SPC No. 494 (28-foot, 10-ton) for Lot "C." Cars No. 358 and No. 494 were ex-ORR (No. 358, ex-ORR 152; No. 494, ex-ORR 196), and were transferred to the SPC in 1899. One source also shows SPC No. 2, a 10-ton, 24-foot car, as also sold to the B&P, but this is not confirmed. The flatcars were 28-foot, 10- and 15-ton capacity, many second- and third-generation SPC/Carter Brothers. Cars No. 101, No. 297 and No. 325 were ex-ORR cars transferred to the SPC in 1894. Cars No. 893 and No. 899 were ex-Pacific Improvement Company. The other known SPC car numbers involved in these lots include:

39 41 55 103 109 237 351 417 419 423 425 445 465 469 481 489 499 517 549
553 581 595 597 607 667 687 689 691 729 759 789 807 817 821 839 853 863

The passenger car was SPC combine No. 18. This monitored-roof car was built by Carter Brothers and was in service on the SPC by 1883. The sale records indicate that car was relettered for Butte & Plumas No. 1. Evidence indicates the fascia boards were lettered "Butte & Plumas Railway," with the other corporate name, "The Truckee Lumber Company," being spelled out under the windows.

Known costs were $670 for the three boxcars in Lot "A," $7245 total for Lot "C" of which $1000 for the combine was given as a note to the SP by the Truckee Lumber Company.

The combine was returned to the Southern Pacific "Laid Aside List" in January 1912, probably because of defaulting on the note. This car was sold again April 17, 1912 to the San Bernardino & Redlands (although lettered and numbered as "Southern Pacific #4"), a narrow gauge line running between the two named towns. The SB&R equipment was absorbed into the "new" Pacific Electric on April 24, 1912, and the line was abandoned in 1916. A fascinating possibility remains that the car could have been photographed with some of the new P.E. steel interurbans, although nothing has come to light to date. The ultimate fate of the old former South Pacific Coast/Butte and Plumas Railway/Southern Pacific/SB&R combine is unclear.

It is assumed much of the TLC equipment was obtained by Swayne and possibly others, between 1912 and 1917.

The Swayne Lumber Company, 1909-1917

There is very little documentary or photographic information on the pre-1917 Swayne equipment. What little exists indicates that the company operated a number of 28-foot logging flats, two of which are definitely known to be ex-Northwestern Pacific No. 5470 and No. 5471. One photo shows what appears to be disconnected trucks, but nothing else is known. It would be a fair assumption that some of the idled B&P equipment may have been used prior to 1917.

With the acquiring of the TLC property, Swayne made a decision to change over from logging flats to skeleton log cars. Perhaps they were influenced by the skeletons owned by nearby outfits like LaMoine and M. A. Burns. They were designed with the help of Seattle Car & Foundry.*

What equipment Swayne had was brought into the new operations.

The Swayne Lumber Company, Oroville, 1917-1940

While a fair amount of equipment was brought into the new operation, much of it was old obsolete or worn out. The following roster includes cars probably not present after 1930, and certainly not on the property at the end in 1940.

**See separate discussion on log cars.*

THE SWAYNE ROLLING STOCK ROSTER

• 1917 – 1940 •

Car No.	Type	Builder	Date	Length	Remarks
1–100	Skeleton log car	Swayne	1917/on	38 feet	*See Below*
1	Oil tank car	?			Capacity unknown.
2	Oil tank car	Swayne		28 feet	Capacity 3070 gals.
3	Oil tank car	Swayne		28 feet	Capacity 3770 gals.
	Gasoline & Diesel	Swayne			Separate fuel tank car tanks unknown.
	Fire fighting car	Swayne		22 feet	Capacity 2532 gals; welded tank, pump, hose reel & upper platform
2, 3	Flatcars	Hammond (?)		24 feet	ex-B&P Ry.
4, 5	Flatcars	Carter Bros.		28 feet	ex-NWP No. 5470 & No. 5471
	Moving car(s)	Unkown		40 feet	ex-SP standard gauge (3 cars)
	Moving car	Swayne		36 feet	the "Bull Car"
	4-wheel dump	WWS or PC&F ?			5 cars total
	Boxcar	Swayne		24 feet	Built on flat; used for commissary.
	Water cars(s)	Unknown		Unknown	No data.
	Block cars			24 feet (?)	2 cars
	Horse cars	Carter Bros.		24 feet	*See Below* (2 cars)
1	Caboose	Swayne		24 feet	*See Below*
2	Caboose	Swayne		24 feet	*See Below*
3	Caboose	Swayne		24 feet	Vertical siding.
	Rotary Snowplow	Swayne	1923	24 feet	*See Below*
	Steam Shovel			24 feet	On cat tracks.
	Steam Derrick				On RR trucks.

• NOTES •

The log cars. *(See separate discussion.)*

The tank cars were homemade, using old boilers and old Carter or Hammond trucks.

The fire fighting car was built as a result of Forest Service regulations.

The two short flatcars were probably ex-TLC or pre-1917 Swayne.

The two 28-foot flats were ex-NWP as noted.

The three ex-SP moving cars were 1920's style straight-silled steel flats. The only modifications were narrow gauge trucks, couplers and rearranged air hoses. The extreme length and width of these cars is the subject of a humorous story in the text.

The Moving Car was built either by Swayne or by a local shop in Oroville, using a spare pair of PC&F trucks. This particular car was called the "Bull Car." The trucks may have come from log car No. 78 which was wrecked in 1925 by one account. This moving car was on the property in the spring of 1927.

The boxcar was built on an old Carter flat and was used for woods commissary service. It had two small end doors for air circulation.

The water cars are reported in one account with no details.

The horse cars were old Carter Brothers boxcars with the roofs cut off. The cars were used prior to 1925 to move horses from Oroville to the woods and back. The roof removal was more for air circulation rather than for clearance of the horses' heads.

The three cabooses were all similar, except the No. 3 had vertical siding. From available data, the cabooses existed by at least 1917, if not earlier. The Carter Brothers trucks are a clue to the earlier date; the CB trucks are noted for their fragile, limited lifespans. Had the cabooses been built in the 1920's, most likely they would have been equipped with Hammond or rebuilt PC&F/Lovestead types.

The rotary snowplow. *(See separate discussion.)*

The shovel was used for railroad and truck road building and was taken to the woods on the moving cars.

The steam derrick was apparently homemade by Swayne, using an old donkey engine boiler and machinery.

Note: There is a persistent story that Swayne purchased cars and/or parts from the Southern Pacific after the standard-gauging of the Nevada-California-Oregon Railway in 1929. To date, there is no documentation or photographic eivdence to support this claim.

• DISPOSITIONS •

When the Swayne Lumber Company shut down, most of the equipment reverted to the Southern Pacific for reasons discussed in the text. Most of the equipment was placed on sale on a bid basis. On September 28, 1939 and lasting until the 30th, an employee of the West Side Lumber Company, Claude Follett, surveyed the 99 log cars, as well as the tank cars, flats, moving car, boxcar, fire fighting car, etc., at Oroville. Not appearing in this survey are the two cabooses and the Shay which were to be sold to WSLC, and it is speculated these had already been surveyed or spoken for by the Tuolumne people. Oddly, the third caboose, the one with vertical siding and which was present in 1939, does not appear in any report or survey and its ultimate fate is uncertain.

The West Side Lumber Company purchased the 99 log cars and Shay No. 6 on one bid from the Southern Pacific. Their bid or offer was for $125 per log car and $1500 for the Shay, F.O.B. Oakdale (the SP connection to the Sierra Railway), which meant they were asking the SP to haul the equipment all the way down the Central Valley at the big railroad's own expense. The offer was accompanied by a 10% deposit check for $1,387.50. The offer was duly accepted by the SP and in February of 1940 the equipment was shipped to Oakdale. The West Side ledgers show freight charges of $1,895.79, and while not further identified, may be the Sierra Railway charges for hauling the cars and locomotive on to Tuolumne.

West Side also purchased the moving car, cabooses 1 and 2 and possibly some or all of the little dump cars, all on separate bids. Costs for these cars are unknown at this time, but a West Side ledger notation for February 22, 1940 shows an equipment expense involving the Swayne purchases for $206 and a freight charge for $303.87 which, while also not identified, may be the 36-foot steel moving car. For the Shay, log cars and the one unidentified car, West Side spent $16,696.91, including freight. What they got was essentially a new railroad "train set," and a radical change in their railroad operation from 24-foot flatcars to 38-foot skeleton log cars.

The West Side considerably rebuilt the two cabooses, Swayne No. 1 becoming (2nd) West Side No. 1 and Swayne No. 2 becoming (2nd) West Side No. 2. The No. 1 survives to this day and is now owned by David W. Braun in Sacramento, California. The No. 2 was turned over on its side at Deadwood Siding in 1962 and no longer exists. The moving car acquired a number (4), which it apparently never had on Swayne, having been referred to as the "Bull Car." The car survived until 1986 when it was turned over to salvage the trucks and brake cylinder. The body was cut up in January 1987.

The rotary snowplow was purchased by the Nevada County Narrow Gauge from the Southern Pacific and shipped to Colfax in October of 1940. The plow was scrapped in 1943 when the NCNG was torn up, having never been used to clear snow.

Of the remaining equipment, the boxcar, flatcars, fire fighting car, etc., were used on the Swayne scrapping train. Thereafter in 1940, all remaining equipment was scrapped at Oroville under ownership of the Southern Pacific, except for the fire fighting car and tank car No. 3. In March of 1940 these two cars ran away from the scrapping train and were completely destroyed when they left the rails at a road crossing near Whitewashed Trees.

The Swayne Skeleton Log Cars

Swayne did not own skeleton log cars prior to 1917, at least of the type in use after that date. While the Truckee Lumber Company had used Russell-type skeletons, it is not known whether Swayne had similar cars, or used the TLC cars at all. Two other Northern California loggers, LaMoine and M. A. Burns had skeleton cars, and it may be that Swayne was influenced by their equipment. It is known that Swayne decided, just prior to acquiring the TLC properties, that the longer haul would mean three-truck Shays and bigger cars would be needed to improve efficiency in the form of total board feet per train, since smaller locomotives and flatcars would not allow for any additional increase in load.

By March 24th, 1917, the Seattle Car & Foundry Company had completed their bunk part drawings for the new Swayne cars, using a design based on similar cars built for M. A. Burns and LaMoine Lumber & Trading. These earlier cars used bunks made from a combination of riveted plates, channels and I-beams. Also, each used three castings per bunk, consisting of bearing pads and the bolster truck pivot plate. The M. A. Burns cars were shipped in April 1914 to Castella, California. These were 32 feet long and equipped with KC airbrakes and link-and-pin couplers. The truck's upper bolster was made from wood, as was the wood filler piece on top of the bunks.

In February 1916 the LaMoine cars were built and shipped. The eight cars had improvements over the Burns cars in that they were longer (34 feet, 9 inches) which gave them larger capacity (5M). Additionally, they had new 52-inch w.b. all-steel trucks, plus straight and automatic air brakes. Interestingly, these cars later went to Burns to join the others, operating under the new Castle Crag name.

While the Swayne cars generally followed the previous design, there were two important differences: the trucks, while ostensibly SC&F, were of the outside-hung brake style. In actuality only the bare journals, boxes, wheels and springs were made and shipped from SC&F's Portland plant (not the main plant in Tacoma, Washington) with the actual fabrication done by Swayne at Oroville. The brake system was single train line only.

An order for fifty sets of bunks was placed April 12, 1917. The cars were assembled in Oroville, with Swayne furnishing the wooden sills, truck side frame straps, brake gear and all labor. The Swayne skeleton log cars were a unique design, almost completely dissimilar from any other narrow gauge log car "connected truck" type general arrangement.

Parenthetically, it should be noted that there were several close cousins to the Swayne cars, although they did not operate in the United States. The Granby Consolidated Mining, Smelting and Power Company of Anyox, British Columbia, Canada, ordered ten sets of parts for "two-bunk, connected trucks" in January of 1924. The parts were shipped to British Columbia, and eight cars were constructed at Anyox and two more at Alice Arm. This equipment was used in logging operations at Alice Arm and Kitsault Flats, B.C., from March 1924 to August 1927. Additionally, two other cars were operated by Brooks, Bidlake, Whithall Ltd. of Lang Bay, B.C., east of Powell River, from 1920 until 1923. Length of the cars is unknown but is estimated to between 36 and 38 feet overall. These cars differ from the Swayne cars primarily in the arrangement of the brake system and, most noticeably, were equipped with the standard SC&F/PC&F 52-inch wheelbase all-steel trucks, as oposed to the trucks used on the Swayne cars which were, of course, unique. Identical to Swayne's method, the Alice Arm cars did not use chains to anchor the loads.

SC&F/PC&F had a stock off-the-shelf narrow gauge skeleton log car design — the 36-foot four-bunk style — in production at this time. West Side Lumber Company and the Diamond & Caldor Railway both purchased cars in April and May of 1917, and it is not known if the Swayne people considered this design over the two-bunk fabricated style at all. Ultimately, the D&C had 84 of the four-bunk cars, while the West Side never followed up on its twelve-car trial purchase. In fact, after the arrival of the Swayne cars on the West Side, the surviving eleven four-bunk cars were dismantled and the parts plus extra trucks and bunks were used to create twelve two-bunk "copies," lengthened to 38 feet.

If the engineering and sales departments of Pacific Car and Foundry could have prevailed upon Swayne, the appearance of their connected-truck log cars would have been like the cars shown in this photograph. The log train is at Kitsault Flats, B.C., being operated by a gypo contracting to Granby Consolidated Mining, Smelting and Power Company. By the spring of 1919, all design work and tooling was in place to make narrow gauge versions of the standard PC&F 42-foot connected-truck cars. As mentioned, Swayne purchased only some parts and constructed the cars themselves, complete with homemade trucks, at the Oroville mill. (J. B. Scott photograph, British Columbia Forest Service collection)

When Swayne decided to change to connected-truck log cars, Seattle Car & Foundry was able to point to two nearby customers who used this general arrangement: M. A. Burns and LaMoine Lumber and Trading Company. The design used a riveted bunk arrangement and new 52-inch WB diamond archbar trucks. In these two photographs, taken about 1919 on LaMoine, the design details for these cars can be seen.

The bunk assembly is made from individual 3/8-inch steel plates riveted together. The truck pivot bearing plate is the standard SC&F casting used on various two- and four-bunk connected-truck designs. In the foreground of the one photograph is the standard SC&F 52-inch WB truck. Note the odd mounting arrangement of the triple valve to the side of the sill, rather than to the air reservoir. The 1917 Swayne cars differed in some aspects from these cars, which were shipped to LaMoine in February 1916. Length was increased from 34 feet, 9 inches to 38 feet. The initial Swayne design was automatic air only and, as discussed, Swayne elected to build trucks of their own design from parts furnished by SC&F's Portland, Oregon plant. The bunk length was the same as LaMoine (7 feet, 6 inches) but in March 1919, as a part of discussions between Swayne and the new PC&F over the problems with the rivet bunk design, a general arrangement drawing was prepared showing full eight-foot bunks and considerable reinforcement of the riveted plates. It is not known if any heavier bunks were built, but they would have been obsolete in any event, as Swayne began ordering new cast-steel bunk supports in April 1919.

This narrow gauge connected-truck design was not popular and only five customers purchased cars or parts to construct such cars from 1914 to 1924, and only 126 cars were built to the general arrangement during that time. (Photos: Maylon "Sandy" Coots collection)

The four-bunk log cars proved very popular with logging outfits. Among others, they were used by Diamond Match, Biles-Coleman, M. J. Scanlon and the Oregon Lumber Company, along with the aforementioned Diamond & Caldor Railway (The California Door Company), which proved to be the biggest customer for this design. The narrow gauge two-bunk design for some reason had limited appeal, and only the five customers cited (three in the United States and two in British Columbia) ordered cars or parts following the general arrangement of precursor M. A. Burns design.

Perhaps this was because there were a number of inherent problems with the fabricated bunk design. For SC&F/PC&F, the assembly of these bunks was material- and labor-intensive. For Swayne the riveted construction apparently did not hold up too well in logging service and the bunks began to fall apart. While welding technology was available by this time, it is not known if Swayne tried any field repairs. However, they must have complained to Pacific Car and Foundry Company, because on March 27, 1919 PC&F issued drawing #G2082, covering part number GS-2082 for a one-piece, cast-steel bunk bracket. Swayne immediately ordered 50 sets of the new bracket, apparently as a trial, on April 12, 1919. This was followed by subsequent orders later in 1919 and 1920. Approximately 75 sets total were ordered, and many cars were either changed over or built new with the new design. Some riveted bunk cars lasted until the end in 1939.

The changeover to straight and automatic air brakes occurred in the 1920's. This involved a cast valve body being bolted between the air tank and the triple valve, as well as the addition of a second branch pipe, water trap, second train line and two air hoses at each end of the car. The straight air hose assemblies are smaller, involving three-quarter-inch nominal pipe, in contrast to the 1¼-inch pipe size of the automatic air. Additionally, the straight air hose did not have an angle cock, but rather a standard WABCO shut-off valve located in the train line about 18 inches to the rear of the hoses.

The cars were numbered consecutively from 1 to 100. In 1925 car number 78 was lost in an accident and was not replaced, its trucks being used on the moving car. The cars received an extraordinary amount of maintenance and were kept clean of slash and debris. Starting in 1935 and extending into 1939, 53 cars were extensively rebuilt, and the 1939 Follett Survey found only 16 cars (ones not having been in the rebuilding program) that could be considered to be "fair" or "not too good." The cars were "painted," if that is the term, in a number of different styles. Some cars were completely unpainted, others were painted in "Swayne Red," and yet others had the ends painted and the center sections left unpainted. In the center of the cars a black-painted rectangle held the car number, in "Railroad Roman," and the phrase "38 ft" in block-style letters. The reason for the fresh black rectangle was apparently to have the car number show up prominently, probably for noting board feet per car in the woods.

Most skeleton log car wheels bore "wheel marks": a white or aluminum splotch of paint on the wheel center, just inside the tire area. The trainmen could tell at a glance whether an axle was locked and thereby prevent damage to the journals or wheels. These wheel marks show up very well in a 1940 movie, taken on the West Side, of the newly arrived No. 6 Shay and log cars. Interestingly, many of the surviving skeletons still have traces of these marks. West Side also applied the marks to some of the other equipment. One example, Caboose #1, has the remains of these marks to this day.

Upon arrival of the cars in Tuolumne (West Side's headquarters), they were immediately put into service. Apparently they ran several seasons, until after 1945, in Swayne color and numbers, before being repainted with the more familiar WSLC numbers, capacities and safety stripes (albeit through several years of development, with the safety stripes or half chevrons being added only for the 1960 season). Much of the existing West Side equipment was modified with a second train line to permit the cars to work together. As a minor note, the Swayne cars had elliptical washers on their sides, West Side gradually replaced them with malleables.

As far as is known, the renumbering from Swayne to WSLC was to the same number. Car No. 78, which was destroyed on the Swayne, came back to life in a way, when WSLC constructed a copy in July 1942, seventeen years after the original car ceased to exist. Interestingly, the new #78 did not have the PC&F trucks but an old pair of rebuilt donkey trucks off a wooden moving car. West Side went on to make other copies, numbering them from 113 to 170; car numbers 101 through 112 being occupied by the ex-four bunk cars from PC&F.

Beyond the end of West Side, a number of the former Swayne skeleton log cars still survive: among them, numbers 3, 8, 11, 21, 49, 56, 64 and 70, located in various locations ranging from California to Colorado, and as far east as Green Bay, Wisconsin. All in all, quite a remarkable record and story . . . 1917 to 1991.

The majority of these photos, with the exception of the Swayne truck closeups by author Paul Beckstrom, were taken by West Side Lumber's car shop foreman Claude Folett during a survey of Swayne rolling stock on September 28, 1939. The details of the Pacific Car & Foundry cast-steel log bunks and brackets, plus the two different "styles" of Swayne trucks can be seen. One truck is equipped with the "tall" I-beam bolster, the other with the "short" version. On the photo above, note the painted ends, unpainted center beams and painted area behind the car number. (Folett photos courtesy Russ Simpson collection)

Swayne skeletons as delivered to West Side Lumber Company at Tuolumne, March 30, 1940. Cars have yet to be turned so that "bent" cheese blocks all line up on the brow log side for dumping logs. Painted area behind car number and elliptical washers on main beams are clearly evident. (Nick Bailey collection)

This series of closeup photographs show the various details of the Swayne skeleton log cars. This includes the various wheelmarks and the substitution of malleable washers by West Side during gradual rebuilding. (Authors' collection)

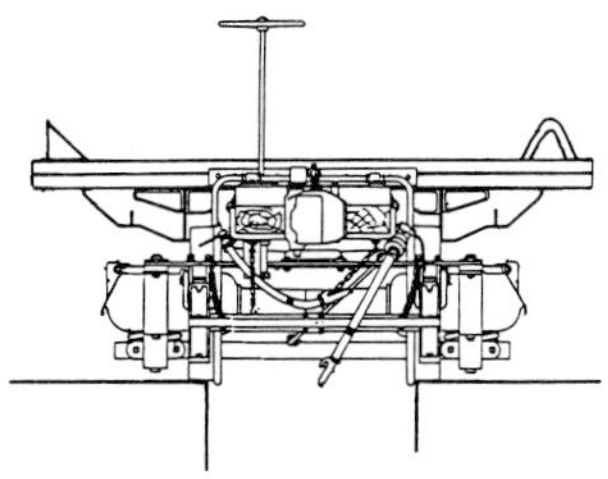

Swayne Lumber Company 38-Foot Skeleton Log Car

Scale: 3/16 inch = 1 foot (Full size Sn3 scale)

Plan by David W. Braun

Color: Unpainted natural wood — most cars.
Ends only — Oxide Red — some cars.
Entire car — Oxide Red — some cars.
Bunks/Trucks — Natural Iron.

The Swayne Cabooses

Based on photographs, the Swayne cabooses existed by 1917. Unlike common carriers, cabooses on logging railroads were not standard, although there were exceptions to this general rule. In the Oroville area, the Hutchinson Lumber Company (Feather River Pine Mills) operation at Feather Falls used cabooses, while other nearby logging outfits such as Diamond Match (standard and meter gauge operations), narrow gauge loggers LaMoine, M. A. Burns, Castle Crag, Feather River Lumber and Spanish Peak, did not.

The Swayne cabooses were built for practicality. The long run to the Merrimac district, with longer trains, demanded five-man crews and five men could not possibly keep control over an 18- or 22-car train from the cab and tank of a three-truck Shay. In keeping with standard railroad practice, three men (engineer, fireman, brakeman) were aboard the road engine, while the conductor and rear-end brakeman rode the caboose. As noted in the text, this was the usual practice for the MTy turns, with the two brakemen riding the loads on the way back.

The Swayne cabooses were built in response to operating conditions present on the old Butte & Plumas Railway, and in the interest of train control. The comfort or safety of the crews was most probably secondary.

Of the three cabooses, there is only one survivor; yet analysis of this car, plus photographs of all three on Swayne and two on West Side, allows a discussion of the cars in detail.

It is likely the cabooses were built prior to the time of the Swayne-TLC consolidation. The Carter Brothers trucks are a clue to the probable building date: had they been built in the 1920's, most likely the trucks would have been Hammond or PC&F/Lovestead/home-brew types. Although the Carter swing-motion trucks were not known to be durable, they lasted under these cabooses from 1917 to 1940 due to the light duty and Swayne's extraordinary maintenance policy. However, the West Side, in the spring of 1940, took one look at these probable fifty- or sixty-year-old trucks and immediately replaced them with rebuilds.

The carbodies themselves were new from the ground up, rather than a rehashed flatcars or boxcars. Four heavy main sills on the frame, with typical body framing, truss rods and double-sheathing, all added up to a quality job. The interiors had lockers, bench seats, a coal stove, plus a cupola which put the conductor approximately 13 feet up in the air for good visibility. Windows on the Swayne version were somewhat restricted as far as total glass area; the West Side made the cupola virtually all-windows for better visibility. The side cupola windows on the Swayne version were diagonal from each other so the conductor could sit by an open window with his feet stretched out in front, regardless of which direction the caboose was facing in relation to the train.

Caboose No. 1 became West Side (2nd) No. 1, and survived. It underwent at least two rebuildings on the West Side and is owned today by David W. Braun. No. 2 became West Side (2nd) No. 2 and was rebuilt extensively; even more so than No. 1. The rebuilding of the No. 2 occurred after an accident in 1949, when a log from a shifting load ran through the carbody, almost the length of the car. Luckily there were only minor injuries to the two West Side crewmen aboard, but the No. 2 was largely destroyed. Interestingly, at the time of the accident, the No. 2 still bore the Swayne railroad Roman-style numeral "2." In 1962, while the caboose was stored at Deadwood Siding, it was tipped over by persons unknown. Over the years the remains of the car were gradually reduced to only a few pieces.

No. 3, the unique vertical-sided model, was on the property at the end in 1939, but does not appear in any sale list. Its ultimate fate is unknown at this time. In addition to the vertical siding, the No. 3 caboose had windows in the doors, which the other two did not.

Caboose No. 1 closed out the narrow gauge logging era on the Pacific Coast when it was used on the last westbound train movement on the West Side, June 7, 1961.

The significant point about the Swayne cabooses is their long lifespan — 1917 to 1961 — working for two narrow gauge logging railroads. Who knows how many thousands of miles they accumulated on their runs to the Merrimac Forest, Granite Basin, the Little North Fork, Buffalo, Hanlon Junction and Camp 45?

Plan by David W. Braun

Swayne Lumber Company Caboose No. 1
Scale: 3/16 inch = 1 foot (Full size Sn3 scale)

At right: Caboose No. 1 at Oroville in 1939, exhibiting a fresh coat of paint and even fresher "snooze stripes" emanating from the cupola. (Bert Ward, Nick Bailey collection) Below: Least photographed side of No. 1 at West Side in early 1940's. Cupola window on this side was diagonal from one on the other, so the conductor could ride with his feet outstretched and his head at the window when traveling cupola-end-first. (J. E. Boynton)

Swayne cabooses No. 1 and No. 2 at Oroville, February 12, 1940, awaiting shipment to West Side Lumber at Tuolumne. (Al Phelps)

After the arrival of the former Swayne cabooses in Tuolumne, new owner West Side immediately changed out the old Carter Brothers trucks. In this photograph, taken in the summer of 1940, caboose No. 1 shows off its "new" trucks. These trucks were not much newer, as they were used on West Side as early as 1904. Rebuilding is evident by the mix of Lovsted and Hammond truck journals and the four- and nine-spring bolster arrangement. West Side also added coupler lift bars, not used by Swayne. Other Swayne features remaining at this time: cupola, position of stove, brake cylinder, and Roman-style "1." It is evident that West Side crews shared the same vice as their Swayne counterparts, as the stains from "chew" are almost identical to the view on the previous page. (John E. Robinson collection)

The photograph below is thought to have been taken in the late 1940's. The transformation of the old Swayne caboose to its final West Side appearance is almost complete, save for the Roman-style "1." West Side has added a second train line and also changed the trucks again. The removal of one spring per side was an attempt at a smoother ride but the result was sagging trucks. The caboose presented this appearance for the rest of its life, except for the final season when it sported a new cupola. (David W. Braun collection)

Swayne caboose No. 2 at Oroville in December 1937. (William Darrough photo; Paul Beckstrom collection)

Caboose No. 2 at Oroville, June 2, 1938. Swayne ran its cabooses with automatic air only. Like the No. 1, the brake cylinder and Carter swing-motion trucks were removed once the caboose arrived at West Side. (Nick Bailey collection)

West Side caboose No. 2 in camp, early 1940's. The platform is being used to transport some roll roofing and one extra gladhand. (J. E. Boynton)

Caboose No. 2 rolls into the Tuolumne millyard, still with its original cupola. (J. E. Boynton)

The Swayne Rotary Snowplow

The Swayne Rotary Snowplow acquired a post-Swayne lease on life, thanks to the Nevada County Narrow Gauge Railroad in Grass Valley, California. Interest in the little machine seems to be in inverse proportion to its size.

The snowplow began life as a locomotive — an 18-ton Class A Climax, constructed sometime prior to 1903. The precise date is unknown at this time, as is the construction number (believed to fall somewhere between #373 and #473). It is known the engine was built with a "T" boiler, round water tank and typical Class A Climax extended cab and roof. The truck wheelbase was three feet ten inches, with 28-inch diameter drivers.

The whereabouts of the locomotive from 1903 to 1909 is not known. Swayne obtained it in 1909 and used it through 1916. As noted in the locomotive roster, the road number is also unknown. It might have been "1" or it might have been numbered with its Climax construction number, a practice that was not uncommon. When Swayne did acquire this locomotive, it got a rare one. While Class A Climaxes were fairly common back east, only about four are known to have been operated in the State of California.

While the Class A Climax was designed essentially as a "throw-away" tramway locomotive, the heavy wooden construction tended to make them long-lasting, and Class A Climax engines with a half-dozen owners over the years were quite common. When a Class A Climax was set aside, as the Swayne engine was in 1916, it was probably for some mechanical reason (boiler, engine or trucks) rather than problems with its design. It has been speculated that the Swayne locomotive was set aside in 1916 because the wheelsets were worn down and the marine engine had developed a serious problem.

Class A Climax wheelsets were castings and did not use tires. To replace a wheelset would have meant ordering a new set from Climax in Seattle or Corry, Pennsylvania, or by turning the existing casting down and trying to fit a tire over it.

In 1923, when Swayne decided to build a snowplow, the latter was done, accounting for the new wheel diameter of thirty inches and the strange look of the wheels with their thick-looking rims. The two-cylinder Climax engine was replaced with a similar two-cylinder marine unit. The boiler, tank and other fittings were stripped off. The blade and funnel assembly came off a California Highway Department (now called "Caltrans") snowplow truck. The basic Climax locomotive frame and trucks were retained, with the brake beams being removed from the front truck. Steam for the new marine engine was supplied by Swayne locomotives, and all were fitted with a steam line running from the steam dome to the front of the locomotive. The operator's shack was not fitted initially. One eyewitness reported that the unit ran as late as 1929 as a bare flatcar with no protection for the crew.

When the NCNG purchased the plow, it had not been operated by Swayne for a number of years, and the wood sills were rotted in front, causing the funnel to be out of position. According to the NCNG book by Best, the Master Mechanic completely rebuilt the plow after its arrival in October 1940. Thereafter, the plow was stored at Grass Valley. Strangely, it was never used because there was little snow during the remaining winters that the NCNG operated.

The plow was subject of an untold number of photographs and came to the end of its life in 1943 when the Nevada County Narrow Gauge was scrapped.

Rotary snowplow in Nevada County Narrow Gauge livery at Nevada City. (John E. Lewis; Paul Beckstrom collection)

Bert Ward; collection of Mallory Hope Ferrell

Nick Bailey collection

Nick Bailey collection

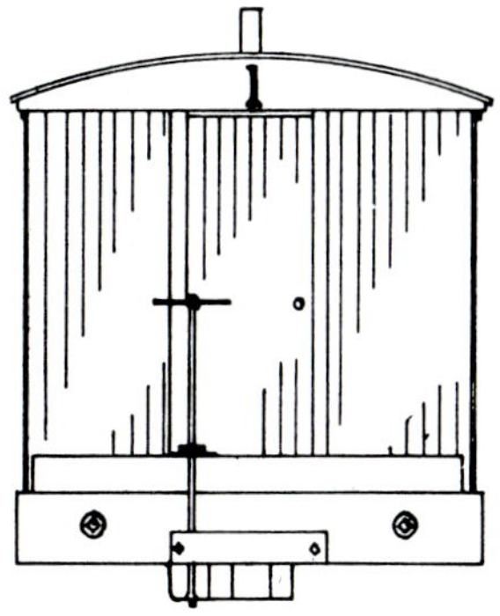

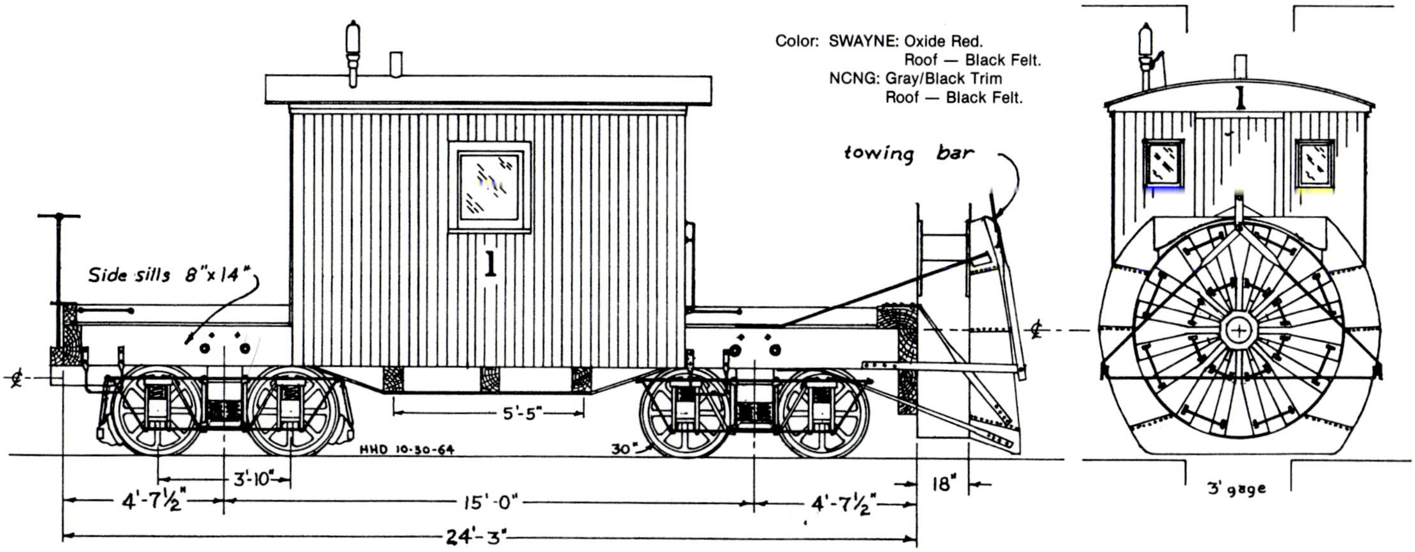

Swayne Lumber Company Rotary Snowplow
Scale: 3/16 inch = 1 foot (Full size Sn3 scale)

Plan by Herman Darr

Caboose No. 3? This partial photo, taken in the deadline during scrapping, shows a caboose with vertical siding and a window in the door. Ex-employees remember this caboose being lettered "B&P Ry." (J. E. Boynton)

Requiem for caboose No. 2. After surviving over forty years of logging operations, No. 2 was deliberately destroyed in 1962 after cessation of West Side at the eerily named place called Deadwood. Shown in 1967, some of the construction details of the Swayne cabooses are evident, including the heavy main frame members and the automatic air lines. The straight air line was added later by WSLCo. The pipe tee seen is not the former branch pipe connection to the cylinder, but rather is the connection for the conductor's valve which was located in the cupola. Nothing remains of this caboose today, save for a few parts in a private collection. (Hart W. Corbett and Nick Bailey collections)

Tank No. 3 at the Oroville mill steam plant. Tank was from an old boiler. (Bert Ward; Nick Bailey collection)

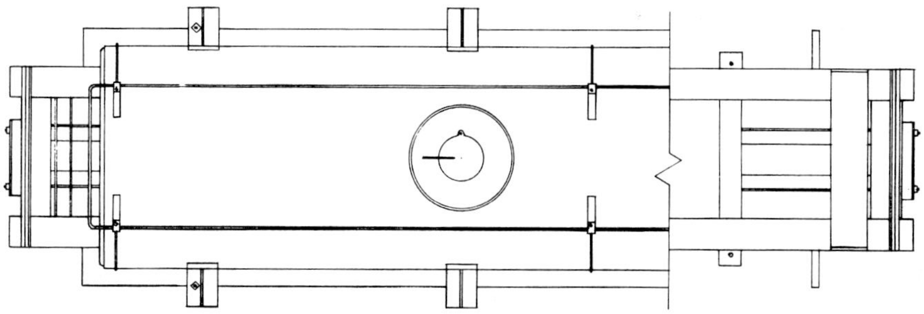

Swayne Lumber Company Tank Car No. 3

Scale: 3/16 inch = 1 foot
(Full size Sn3 scale)
Plan by John E. Robinson

Color: Unpainted wood — Frame.
White/Gray — Tank.
Bunker C Oil Spills — Black
Trucks — Natural Iron.

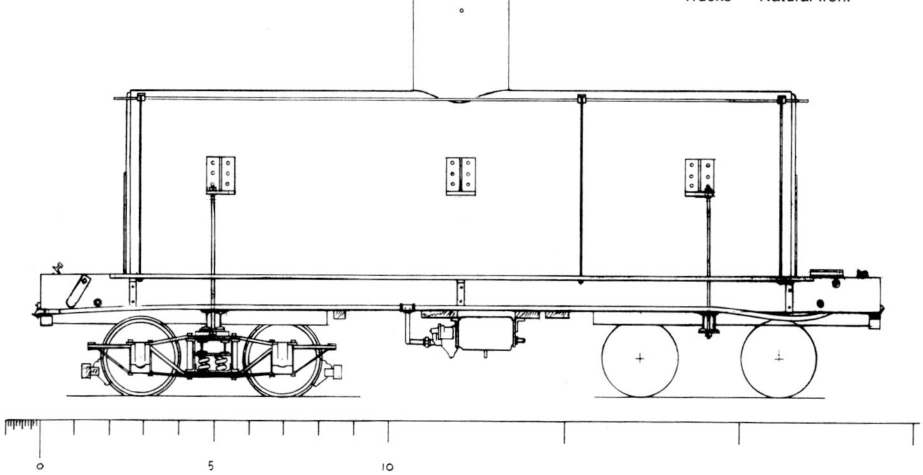

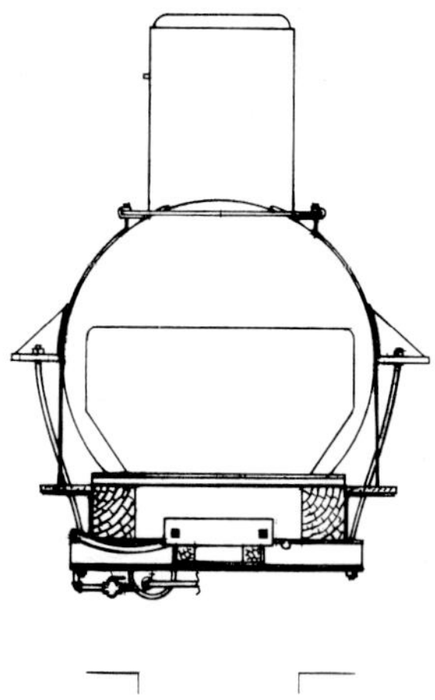

"Bunker C" tank car No. 2 at Oroville, July 2, 1939. (Bert Ward; Nick Bailey collection)

No. 3 eases by the wood warehouse, boxcar in tow, while making up the dismantling train, March 17, 1940. Note Sacramento Northern overhead wires at transfer. (J. E. Boynton)

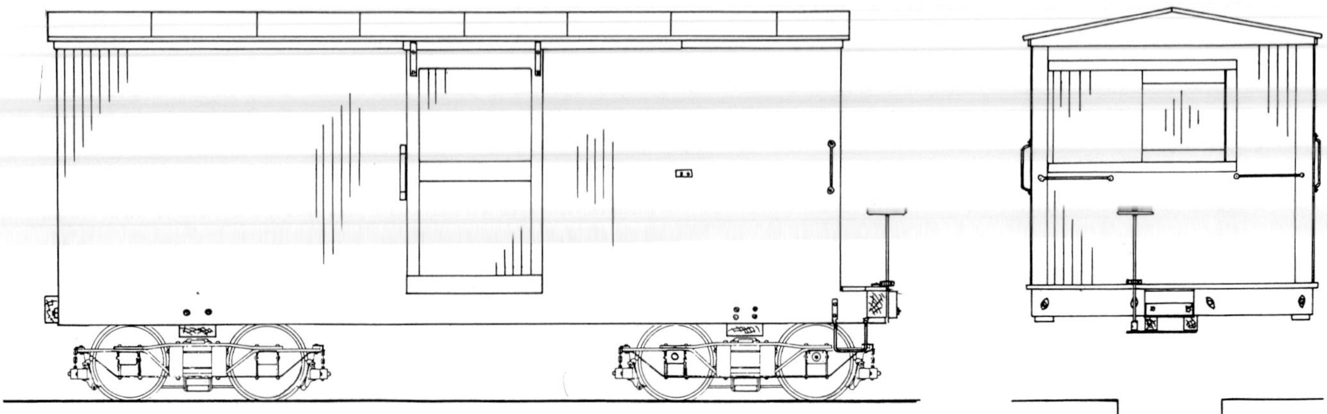

Swayne Lumber Company Boxcar
Scale: 3/16 inch = 1 foot (Full size Sn3 scale)

Color: Swayne Oxide Red
Roof — Black (felt roll roofing).
No needle beams.
Westinghouse "K" brakes.
Carter swing motion trucks.

Plan by David W. Braun

Northwestern Pacific 5400 series 28-foot flatcar, typical of the secondhand equipment purchased from the Southern Pacific's "laid-aside" list by both Truckee and Swayne Lumber companies. (Fred Stindt collection)

Flatcar built by Carter Brothers, circa 1880.
Paint: Box car red; white lettering.
Dimensions taken from existing car. Details from photos.
Cars assigned odd numbers 1-579. Vacated from roster 12/31/09.
Sold to Butte & Plumas Ry. 11/1910.

Side sills: 4½″ × 11″*
Intermediate sills: 4½″ × 7″
Center sills: 4½″ × 7″
End sills: 5½″ × 7″ × 7′ 6″
Truss beams: 4″ × 8″
Bolsters: 4″ × 12″

*4½″ × 10″ side sills on some cars.

Cordwood rack used on some cars.

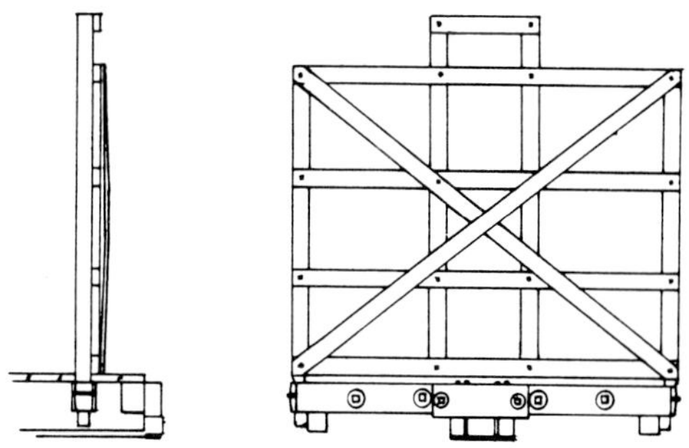

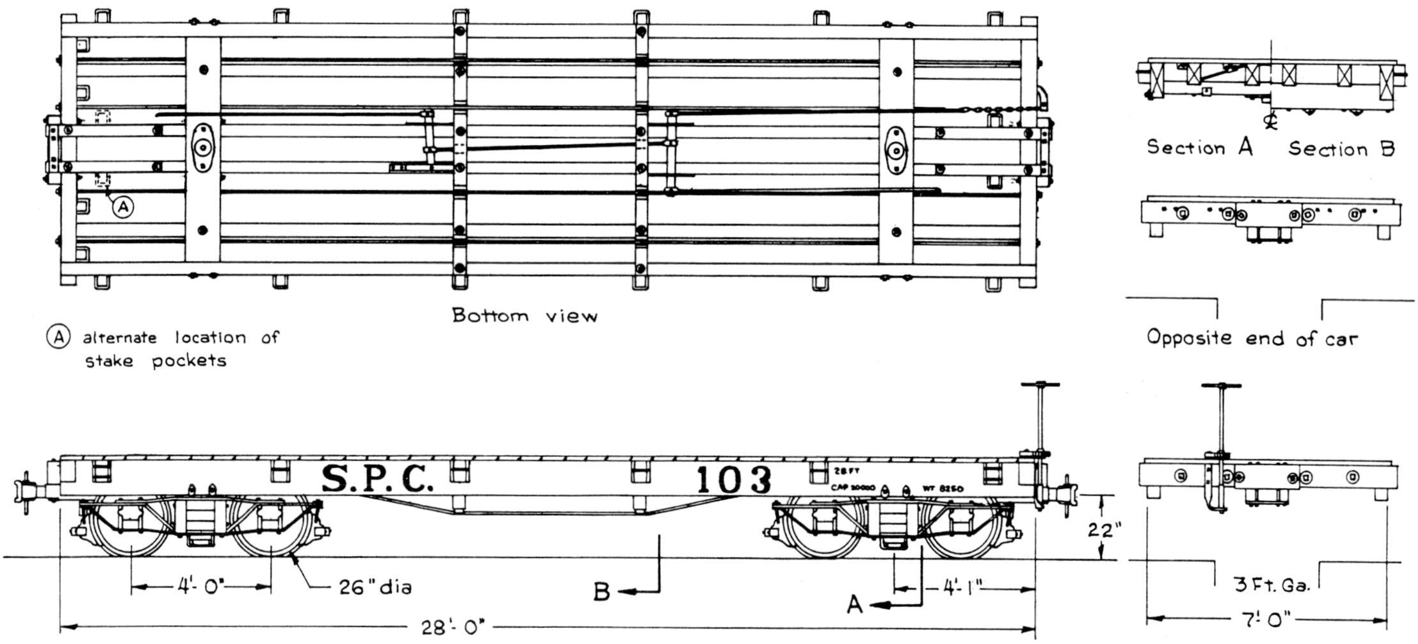

South Pacific Coast RR Flatcar No. 103

Scale: 3/16 inch = 1 foot (Full size Sn3 scale)

Plan by Herman Darr

Swayne Moving Car at Oroville near the end of operations. (Bert Ward)

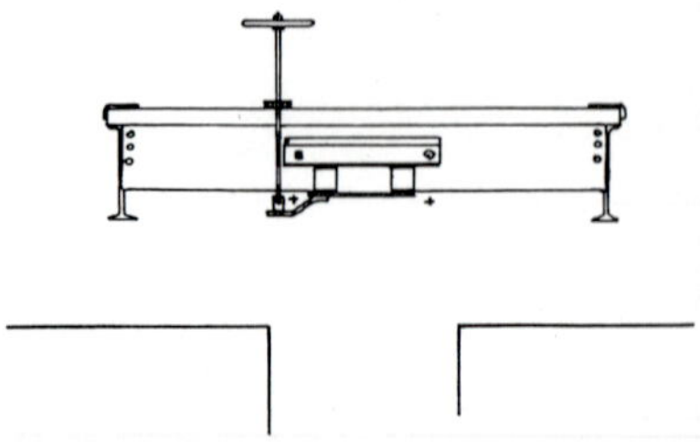

Color: Natural grays
Brakes: Hand and "K" system. Cylinder is bolted to (1) 2x6 and (1) 2x12.

Moving car or "Bull Car," under West Side ownership, at Tuolumne with a substantial load. Date unknown. (Bob Brown)

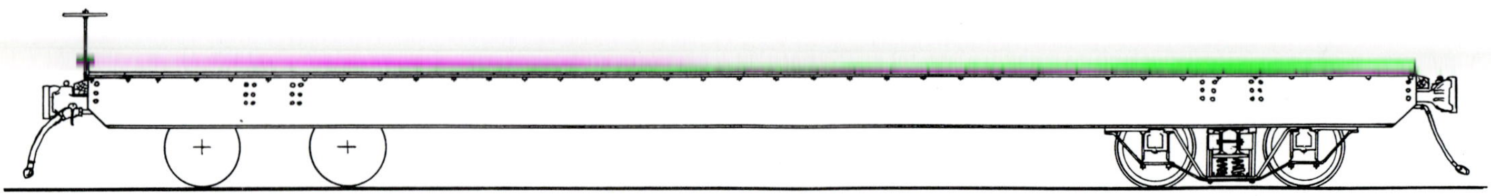

Swayne Lumber Company 36-Foot Heavy-Duty Moving Car

Scale: 3/16 inch = 1 foot (Full size Sn3 scale) Plan by David W. Braun

Above: Internal-combustion, crawler track mounted shovel used for roadbed construction. (Paul Beckstrom collection)

A steam-powered Erie Roadroller is seen at the Oroville mill. (Nick Bailey collection)

The basic 1939 Swayne Roster was compiled by Paul Beckstrom, with additional notes, comment and opinion by David W. Braun.

Thanks to:

—John E. Lewis for SC&F/PC&F and Climax Mfg. Co. data as it relates to Swayne;

—Russ Simpson for the 1939 Claude Follett Survey and West Side ledger pages;

—Joel Segal for the Southern Pacific Laid-Aside List;

—Joe Johannsen for the taper sill flat data;

—"Butch" Slaughter, who was there;

—Kent Stephens for 1930's roster information;

—David Garcia for the SB&R/PE data;

—Herman Darr for data from Record of Equipment Sold, Transferred, Wrecked or Destroyed, Vol. 1, Southern Pacific, Pacific System.

LOCOMOTIVE ROSTERS
Swayne Lumber Company and Associated Firms

TRUCKEE CALIFORNIA AREA / The Truckee Lumber Company, 1867-1909

No. 1	0-4-0T	Porter #699	July 1885	7×12″			NEW
No. 1	0-6-0T	BLW #4036	1876	12×16″	42″	36,000 lb.	USED
No. 2	0-6-0T	BLW #12015	1892	9×14″	30″	24,000 lb.	USED

• NOTES •

No. 1 Porter was purchased new; Porter records show gauge was 42 inches. No record of this locomotive being moved to Oroville in 1909.

No. 1 New to Eureka & Palisade, for Ruby Hill No. 2, "Dewey." Sold, 1893, Pacific Lumber & Wood Co., Clinton and Truckee, Calif., No. 3, "Dewey." Sold, 1901 to Truckee Lumber Company, as part of the Donner & Tahoe Railroad sale. See below for subsequent history. This is the future Swayne "A".

No. 2 New to Pacific Lumber & Wood Co., Clinton and Truckee, Calif. Assigned No. 2, "Tahoe." Sold, 1901 to TLC. See below for subsequent history. This is the future Swayne "B".

Note: PW&LCo's No. 1 was an 0-4-0T (Porter-Bell No. 147 or 148, 1873, 9×12″, 24″, 15,000 lb.) from Sutro Tunnel Co., at Sutro, Nevada. This was an inside-connected, low profile mining loco. This engine did not come to Truckee Lumber Company ownership.

OROVILLE, CALIFORNIA AREA

Note: It is important to keep in mind that the Truckee Lumber Company and the Swayne Lumber Company were independent firms, operating in two different locations in the general Oroville area, prior to 1917.

The Truckee Lumber Company / September 1909 to September 1911
— Operating as "The Butte and Plumas Railway" —

No. 1	0-6-0T	BLW #4036	1876	12×16″	42″		36,000 lb.	ex-PL&WCo. No. 3
No. 2	0-6-0T	BLW #12015	1892	9×14″	30″		24,000 lb.	ex-PL&WCo. No. 2
No. 3	Shay	Lima #2369	9/16/1910	3)10×10″	29″	2T	66,400 lb.	NEW
No. 4	Shay	Lima #2465	7/28/1911	3)11×12″	32″	2T	105,000 lb.	NEW

• NOTES •

Many previous rosters show these locomotives as either Truckee Lumber Company and The Butte and Plumas Railway, or both, leading to some confusion about separate companies, etc. The Butte and Plumas Railway was wholly owned by the Truckee Lumber Company and was organized as a common carrier for reasons discussed in the text. The "Tahoe Lumber Company," listed on some rosters for c/n 2369, is in error, as no such company existed, insofar as a logging railroad operator. The first three locomotives went to Swayne in 1917.

The B&P No. 4 apparently went to West Side Lumber in May 1912.

The Swayne Lumber Company / 1909 to 1917
— At Swayne Hill (nee Pearl Orchard, California) —

No. ?	Climax	Climax # ?	/190(2)	2)7½×7″	28″	2T	36,000 lb.	USED
No. 1	Shay	Lima #				2T		Unknown
No. 2	Shay	Lima #2458	6/17/1911	3)8×8″	27½″	2T	48,600 lb.	NEW
No. 3	Shay	Lima #2926	8/1/1917	3)11×12″	32″	3T	120,000 lb.	NEW

• NOTES •

The Climax was an 18-tonner with round water tank and extended cab. The Swayne road number is unknown at this time, although it has been speculated that it may have been numbered its Climax Manufacturing Company construction number, which at this time is estimated to be between 373 and 473, according to historian John E. Lewis. Mr. Lewis also believes the construction date to be prior to 1903. It is then evident that the Climax was purchased as a secondhand locomotive by Swayne.

The " No. 1" Shay is speculation at this time. However, two points in favor of its probable existence would be: 1 the numbering of c/n 2458 as "2"; and 2 bits and pieces of a Shay locomotive on the property, including a frame and a tank shell (mounted on Swayne No. 4). Two points against the existence of this locomotive would be: 1 if the Climax locomotive was numbered "1"; and 2 a search of the Lima rosters to date has not shown this engine to have been built new for Swayne or acquired secondhand.

The No. 3 Shay was shipped from Lima, Ohio on August 1, 1917 as Swayne Lumber No. 3. It is not known if this Shay operated as Swayne No. 3 for any appreciable amount of time prior to being renumbered to (new) Swayne Lumber Company No. 1 in 1917. However, there is a photograph showing this locomotive numbered "1" on the cab sides and "3" on the headlights, which would indicate the engine probably operated initially as "Swayne Lumber Co. #3."

All locomotives discussed were physically transferred to the post-1917 Swayne operations.

The Swayne Lumber Company / 1917 to 1940
— At Oroville, California —

"A"	0-6-0T	BLW #4036	1876	12×16″	42″		36,000 lb.	ex-TLC No. 1
"B"	0-6-0T	BLW #12015	1892	9×14″	30″		24,000 lb.	ex-TLC No. 2
No. 1	B	Whitcomb #11955	6/1925	Model "MO"	[illegible]			[illegible]
No. 1	Shay	Lima #2926	8/1/1917	3)11×12″	32″	3T	120,000 lb.	ex-Swayne No. 3
No. 2	Shay					2T		ex-Swayne No. 1
No. 2	Shay	Lima #3078	4/24/1920	3)11×12″	32″	3T	120,000 lb.	NEW
No. 3	Shay	Lima #2369	9/16/1910	3)10×10″	29″	2T	66,400 lb.	ex-TLC No. 3
No. 4	Shay	Lima #2458	6/17/1911	3)8×8″	27½″	2T	48,600 lb.	ex-Swayne No. 2
No. 5	Heisler	Stearns #1028	2/18/1899	15×12″	36″	2T	74,000 lb.	ex-WSLC No. 1
No. 6	Shay	Lima #3302	4/21/1926	3)11×12″	32″	3T	133,500 lb.	NEW

"A" The "A" was operated as a mill switcher until the end, still as a wood-burning locomotive. Scrapped by United Commercial Company in South San Francisco in 1941.

"B" The "B" was set aside by 1930. Scrapped by United Commercial Company in South San Francisco in 1941.

No. 1 The Whitcomb #1 was ex-Spanish Peak Lumber Co. #1, Spanish Ranch, California.

Delivered to Quincy Junction, California, July 1925, by Western Pacific RR. Sold to Swayne, February 1936. Although Swayne repainted the locomotive and made some modifications, it was left numbered "1" which probably made for some confusion with the three-truck Shay No. 1. No record after 1940.

No. 1 Lima work order for this locomotive was issued June 5, 1917, using the general arrangement of Card A-16526. Actual construction took from June 12 to July 30, and the engine was shipped August 1, 1917. As shipped, the locomotive had a truss-rod frame and wood cab and was numbered Swayne 3. It was also superheated. To save on shipping weight, the locomotive was shipped less its fire brick. In January of 1928, Swayne ordered a new girder frame, steel cab assembly and other small parts, including some brake gear and a new air compressor bracket for the new cross-compound pump to be installed. Items were shipped late that month and all actual rebuilding was done at Oroville, including installation of piston valves and Worthington "BL" feedwater heater. Locomotive was sold to the Michigan-California Lumber Company in early 1940, and renumbered M-CL (2nd) #1. The Worthington equipment was removed, a back-up light added and other minor changes made. This locomotive was used on the cut lumber haul from South Cable to Camino and was never a woods engine. Locomotive was scrapped at Camino, California, in 1951.

No. 2 (1st). Not confirmed as discussed.

No. 2 The Lima work order for this locomotive was issued February 5, 1920, using the general arrangement of Card 427 B 5001. Actual construction did not start until February 24th and the engine was not completed until April 7th. For unknown reasons, the engine was not shipped until April 24th. As constructed, the locomotive had a wood cab, truss-rod frame, Superheater, and the locomotive had an 8½-inch cross-compound air compressor. Curiously, the engine was shipped with its fire brick installed. In November of 1926, a steel cab, girder frame and other associated parts were ordered and shipped for installation at Oroville. It is not known when the BL heater and piston valves were installed, but a good estimate would be later in 1928 or 1929. Sold to the Michigan-California Lumber Company as (2nd) #3. Some changes made as with No. 1 above, service on M-CL was same as No. 1 above, and locomotive scrapped in Camino, 1951.

No. 3 This locomotive was constructed from Lima Plan 2134. This was a very popular design and at least 42 locomotives were built to this plan, in five different track gauges, between January 1909 and November 1912. The Truckee Lumber Company engine was the 17th built to the plan and was one of only three of the group to be built new as an oil burner. In the absence of primary records, it is probably safe to say that, due to the density of construction (42 locomotives in 47 months), the Truckee Lumber Company engine was ordered largely from stock, and TLC had only to specify the oil-burning equipment.

The Lima repair parts record shows some spare parts ordered for this engine through September of 1911, and then, curiously, an order in January 1913. This might indicate the locomotive was being used by Swayne, or that someone was refurbishing the locomotive for possible use. (The repair records generally do not show the customer name.) Thereafter, no further parts orders were received by Lima until May 1917.

This two-truck Shay had the dubious distinction of heading the scrap trains for both Swayne and the Michigan-Cal "North Side" show. After its sale to M-CL in 1940, the engine was changed from oil to wood-burning, with the appropriate stack change, and was used as a woods switcher. After 1942, the engine was changed back to an oil burner and was used on the lumber run to North Cable. In somewhat of a reversal of the minimalist maintenance policies of the M-CL, a bell was added, something the engine lacked during the last five years or so of Swayne ownership.

Scrapped, Pino Grande, California, 1951.

No. 4 This locomotive was ordered new by Swayne through the agent Norman B. Livermore & Company, and was shipped to Swayne Siding at Blinzig, California. Evidently, the engine was taken up on the incline to get it to the Pearl Orchard operation.

Judging by the Lima Repair parts list for this locomotive, the line shafts and associated parts were either not set up correctly at the factory, or Swayne was rough on locomotives, as the list shows continual orders for shaft parts from 1911 through 1918, including pinions, couplings, bushings, washers and sleeves.

In January 1914 evidently, the locomotive was wrecked, as the repair parts ordered include, among other items, cab appliances, firebox parts and parts for a stack.

This Shay was later standard-gauged by Swayne for unknown reasons. The engine also had a homemade oil tank mounted on top of its original Lima tank as an expedient against completely rebuilding the tank area. When the locomotive was last used or set aside is open to speculation. Certainly it was not used after the woods switcher jobs were abolished, as Swayne had more mill switchers than they could use by that time.

Scrapped in South San Francisco, 1941, by United Commercial Company.

No. 5 This was the 28th locomotive to be constructed by the Stearns Manufacturing Company. It was ordered by West Side Flume and Lumber Company as their first new locomotive and was lettered for their common-carrier "Hetch Hetchy & Yosemite Valleys Railway." Assigned road number 1, it was also named "Henry J. Crocker," in honor of one of the principals. West Side considerably rebuilt the locomotive, altering its original appearance. The locomotive was sent from Tuolumne to Oroville sometime between 1912 and 1917.

It served as a woods switcher throughout the 1920's and early 1930's. In 1935 the late Otto Perry found it in service at the Swayne mill after which it was stored and never used again. Scrapped in South San Francisco, 1941, by United Commercial Company.

No. 6 Lima issued the shop order for this locomotive on March 9, 1926, using the general arrangement of Card 428 C 5055. Actual construction took place from March 11th through April 15th, and the locomotive was shipped

April 21, 1926. This engine was of a stock Lima design known as "60-3," and there were at least three narrow gauge and five standard gauge locomotives constructed to this particular card number. The other two narrow gauge engines were built for the Madera Sugar Pine and the New Mexico Lumber Company. In the case of the Swayne locomotive, the customer requested that the oil pan, steam brake and firebrick be left off the engine. The Swayne and New Mexico locomotives were built new with slide valves, while interestingly, the Madera engine was equipped with piston valves. The method of shipping all three to the West also varied: the Swayne and Madera engine were shipped on flatcars, but the New Mexico engine was sent out on a set of standard gauge shipping trucks, to Montrose, Colorado. The cost to Swayne for their new locomotive, FOB Lima, Ohio, was $21,735.

An amusing incident that occurred during construction of this locomotive was the discovery that someone in the front office had written in the wrong type air compressor on the specifications: that of a small single 9½-inch type instead of the required large 8½-inch cross-compound. In the spirit of paperwork required by large corporations, instead of simply telling the shop foreman to change the pumps, a regular Lima Change Order (2679-E) had to be drawn and routed through channels. It is not known when Swayne changed to the piston-valve arrangement; probably not until 1928 or 1929. After West Side acquired the locomotive, and had some correspondence with Lima about the valves, Lima entered a notation on the locomotives records: "Customer says piston valves on this locomotive, 2-27-40." It is also not known if Swayne came to any final conclusion about installing a Worthington BL feedwater heater on this locomotive. Some preliminary piping was installed for the appliance, but for some unknown reason the main job was not finished.

As far as Lima repair parts, this engine required very little — a bearing plate in March 1929, Zerk fittings for the universal joints in August of 1929, and after the West Side purchase, square shafts in July 1945, and some sleeves in January 1946.

After the locomotive went to West Side Lumber Company, the unique Swayne spotting features — spark arrestor, wooden lockers on the tank, hose reel, etc. — were gradually removed over a period of time. West Side shop crews reported a great deal of trouble with the homemade piston valves; therefore, in its final years of work it was generally kept in the woods with the older numbers 7, 14 and 15, with road engines being West Side numbers 8, 9 and 10. The engine exists today at Silver Plume, Colorado.

The Butte & Plumas No. 4 Shay This locomotive had at least one twin: Lima #2522, built in March 1912, as the Diamond & Caldor Railway #7 (The California Door Company, Diamond Springs, California). Unfortunately, the existing Lima records are missing sheets from the 1911-1912 time period; therefore, we have been unable to determine the Lima Card Number for these two locomotives, or if any additional standard or narrow gauge locomotives were built to the same Card number. The Repair Parts List for both locomotives is also missing. This would help in determining exactly when the Butte & Plumas locomotive was shipped to Tuolumne.

We did not list the B&P #4 on the 1917 Swayne roster because there is little or no evidence that the engine was still on the property at the time of the 1917 reorganization; it may have left Oroville as early as May 1912. We do have the oral statement of a former West Side engineer, Bert Bergstrom, that the locomtive was on the Tuolumne property when he came to work in 1916.

(See the commentary and text for further discussion of this locomotive.)

Hetch Hetchy & Yosemite Valleys RR (West Side Lumber Company) No. 7, ex-Butte & Plumas Ry. No. 4, at Tuolumne in 1920. This photo, the earliest of this locomotive known to exist, was taken eight years after the engine was built. Aside from the HH&YVRR lettering and lack of Lima striping, this is how the engine looked while on the B&P. (Russ Simpson collection)

Commentary on the Locomotive Roster

by DAVID W. BRAUN

The Truckee area operations contributed very little to the new TLC Oroville operations. Apparently, the TLC Porter was not sent to Oroville. Other than the two ex-D&T RR 0-6-0T's, only a few flatcars and rail came up from Truckee.

The Butte & Plumas No. 4 three-truck Shay apparently had little use, as it was shipped from Lima, Ohio on July 28, 1911, and the Truckee Lumber property was shut down in September of that same year. After the operation was suspended, the No. 4 was in a kind of limbo. According to various sources, the locomotive could have left for Tuolumne as early as 1912, or as late as 1917, although one newspaper reported the locomotive as being shipped to Tuolumne in May 1912. The earlier date would tend to be more valid, as a brand-new $18,000 (in 1911 dollars) locomotive was not something to be left standing idle. Where the West Side "investment" in TLC and the Heisler locomotive figure in this "drama" is still to be determined. If the Swayne Lumber Company took possession of this locomotive on paper or physically, it must have been just long enough to be traded or sold off to West Side.

The Climax locomtive must have come to Swayne as a secondhand locomotive. It was a Class "A" and not the Class B model (or something with Class B trucks, as was speculated by historians after seeing it in its second form as a snowplow). The locomotive was worn out and set aside by 1916. The engine was on the property through the 1917 consolidation and in 1923 was converted to the familiar Swayne /NCNG snowplow, with thirty-inch diameter wheelsets. All of the Swayne locomotives were equipped with steam piping to power the plow's engine, which by that time was a marine-type, the original Climax engine having been replaced. The Swayne "No. 1" Shay is pure speculation as outlined above. The existence of this engine can be inferred from both the numbering sequences and from photographic evidence, but unfortunately, there is no Lima construction number available to firmly establish the locomotive's presence at the Oroville operations.

With the acquisition of the former TLC/B&P property and equipment by Swayne in 1917, one would think that the eight locomotives involved would have been renumbered into a unified numbering system. But what actually took place was: the little 0-6-0T's were "unnumbered"; the Butte and Plumas Shay No. 4 was already gone or shortly thereafter traded/sold off; the Climax was set aside; and the Swayne "No. 1" Shay was either set aside or unserviceable. This left only three locomotives of the original eight to go into a new numbered roster. Adding to all the confusion, the West Side Lumber Heisler was in effect the ninth locomotive for the new roster.

It does make sense that brand-new Swayne No. 3, arriving about the time of the consolidation, would be given the first spot on the new roster. What is not clear is why No. 2 was given to a locomotive that did not arrive until the spring of 1920. It may be that No. 2 was occupied from 1917 to 1920 by the Swayne No. 1 Shay (or the Climax, possibly). It does make some sense that road number "2" was occupied by some locomotive, otherwise why would the Heisler, perhaps arriving at the time of the new roster setup, be given road number "5" instead of "2"?

It is not known at this time why the two 0-6-0T's were left off the numbering unless they expected to see little use. If so, Swayne was not a good judge of old engines, as the "A" was used to the end in 1939. The Whitcomb was another exception, as after 1936, Swayne had two No. 1's, in the form of a big three-truck Shay and a sixteen-ton switcher, which may have led to a "Who's On First" situation at times.

A few words are in order regarding Swayne's rebuilding of the three big three-truck Shays. All three were extensively rebuilt. The No. 1's and No. 2's older truss-rod frames were changed out for girder types, purchased by Swayne from Lima. The BL feedwater heaters were purchased from Worthington. Of the conversion of all three from slide to piston valves, while there was an aftermarket kit available from the Baker Marine Division of the Pillord Corporation (this was the kit installed on the sister Madera Sugar Pine locomotive by Lima during its construction in January 1927), the master mechanic at Swayne, Clyde Masteller, elected to make his own patterns and have them cast in Oroville. Interestingly enough, the patterns still exist. The work done on the Shays by Mastellar and his shop crew was done so expertly and the locomotives were so well-maintained, that a search of Lima records for repair parts ordered by either West Side or Michigan-California Lumber after 1940 show only a few small items. MCL managed to break a throttle arm assembly on the former No. 2, and West Side only needed some sleeves and couplings for one of the trucks of the No. 12 (nee "6").

Of the two surviving locomotives, only the big No. 6 (West Side Lumber Company No. 12) could be considered to be a Swayne locomotive. West Side Lumber Company No. 7 (nee Butte & Plumas No. 4) is more of a West Side engine, as it gave that company 43 years of service. The Truckee Lumber Company used it all of two months and Swayne apparently not at all.

Compiled by Paul Beckstrom and David W. Braun

Thanks to:
John E. Lewis for Climax Mfg. Co. data;
Allen Copeland for Whitcomb data;
David Spohr for Pacific Lumber & Wood Co. loco data;
George R. Kadelak for girder frame data and Pillord Corp. data;
Ray Schuck and the Allen County Museum for help with "Order 462";
Staff at California State Railroad Museum Library for help with the Lima "A" and "B" files.

Swayne's 0-6-0T "A," Baldwin built in 1876 (ex-Ruby Hill RR, ex-Pacific Lumber & Wood), was the mill switcher for many years. These views show the venerable old wood-burner performing her chores around the millyard at Oroville during the final years of operation. (Photos at right: Bert Ward; Nick Bailey collection. Middle photo: W. C. Whittaker. Lower photo: Al Phelps)

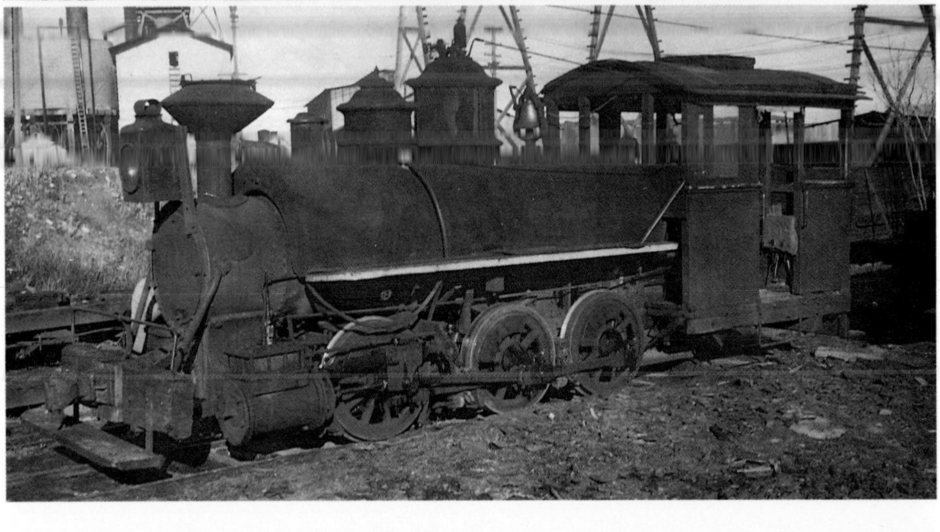

By the late 1930's, Swayne's other 0-6-0T ("B") was in sad shape and reallocated to the deadline at Oroville. Traces of its former owner, Pacific Lumber & Wood, still show on her tank side. This locomotive, along with the "A," was scrapped in 1941. (Above and left: Bert Ward; Nick Bailey collection. Lower photo: W. C. Whittaker)

Whitcomb No. 1 was built in 1925 for the Spanish Peak Lumber Company of Spanish Ranch, California. It was sold to Swayne in 1936 and, although the new owner made some modifications and repainted the locomotive, it still retained its old number "1." As noted, this led to some confusion with Swayne Shay No. 1. At left: The locomotive is shown in August 1925 at Spanish Ranch, just two months old and in its original Caterpillar yellow paint scheme. (Ted Wurm) Below: At the same location in 1930. Note the raised hood due to an apparent motor replacement. (J. E. Boynton collection) Below: On the deadline at Oroville in June 1939. (William Darrough; Paul Beckstrom collection)

Above: Shay No. 1 in its original configuration, alongside the Swayne mill in 1919. As shipped the locomotive had a wooden cab, kerosene lights and slide valves. (J. E. Boynton collection) Below: At Oroville in 1939, after rebuilding. In January 1928, Swayne ordered a new girder frame, steel cab, plus numerous small parts. Rebuilding was done at Oroville and included installation of a Westinghouse cross-compound air pump, Worthington BL feedwater heater and piston valves. (W. C. Whittaker)

Above: Former Swayne No. 1 at Camino, California on June 25, 1940, after
Above: Former Swayne No. 1 at Camino, California on June 25, 1940, after being sold to the Michigan-California Lumber Company. (Al Phelps) At right: A closeup of No. 1, sans-tender. The Swayne-applied Worthington BL model feedwater heater is quite evident. It was subsequently removed by Michigan-Cal. (R. S. Polkinghorn) Lower: Michigan-Cal 2nd No. 1 at South Cable, February 1942. The numeral "1" appears on the locomotive a total of fourteen times. No. 1 was scrapped at Camino in 1951. (Bert Ward collection)

Above: Swayne No. 2 at Oroville, June 1939. Note the piston valve conversion. (J. E. Boynton) At right and below: The ex-Swayne No. 2, now Michigan-Cal 2nd No. 3. The photograph below shows the locomotive after arriving at Camino in June 1940. (R. S. Polkinghorn) Right: At the same location in 1949, two years away from scrapping. (Bert Ward collection)

Above: The No. 3 in the Oroville deadline for winter storage, November 27, 1937. (W. C. Whittaker) Below: Switching the scrap train. The locomotive had no bell during its last few years of operation. (Ted Wurm)

Above: Ex-Swayne No. 2, now Michigan-California 2nd No. 4, switching at North Cable, October 26, 1945. (Al Phelps) Below: The locomotive at Michigan-Cal camp 15 in 1941 as a wood-burner. (Bert Ward)

Swayne Shay No. 4 on the Oroville deadline in the late 1930's. As noted, this locomotive saw little use after the woods switcher jobs were abolished in the early 1930's. The photos at right show the second tender tank (oil) sitting on the lower water tank. Note the Shay frame lying in the weeds in the foreground. (Both photos Paul Beckstrom collection) Above: No. 4 on January 2, 1940. Its homemade oil tank has been removed. (Bert Ward)

Above: Hetch Hetchy & Yosemite Valleys No. 1, the "Henry J. Crocker," before going to Swayne sometime between 1912 and 1917. (Nick Bailey collection) This Heisler served as a woods switcher in the 1920's and early 1930's as the No. 5. At right: The locomotive also saw service at the Swayne mill, but after 1935 it was stored and never used again. (Right: Harold Vollrath collection. Below: Otto Perry)

Swayne Lumber Company Shay No. 6 (Lima 1926)

Drawn by Al Armitage

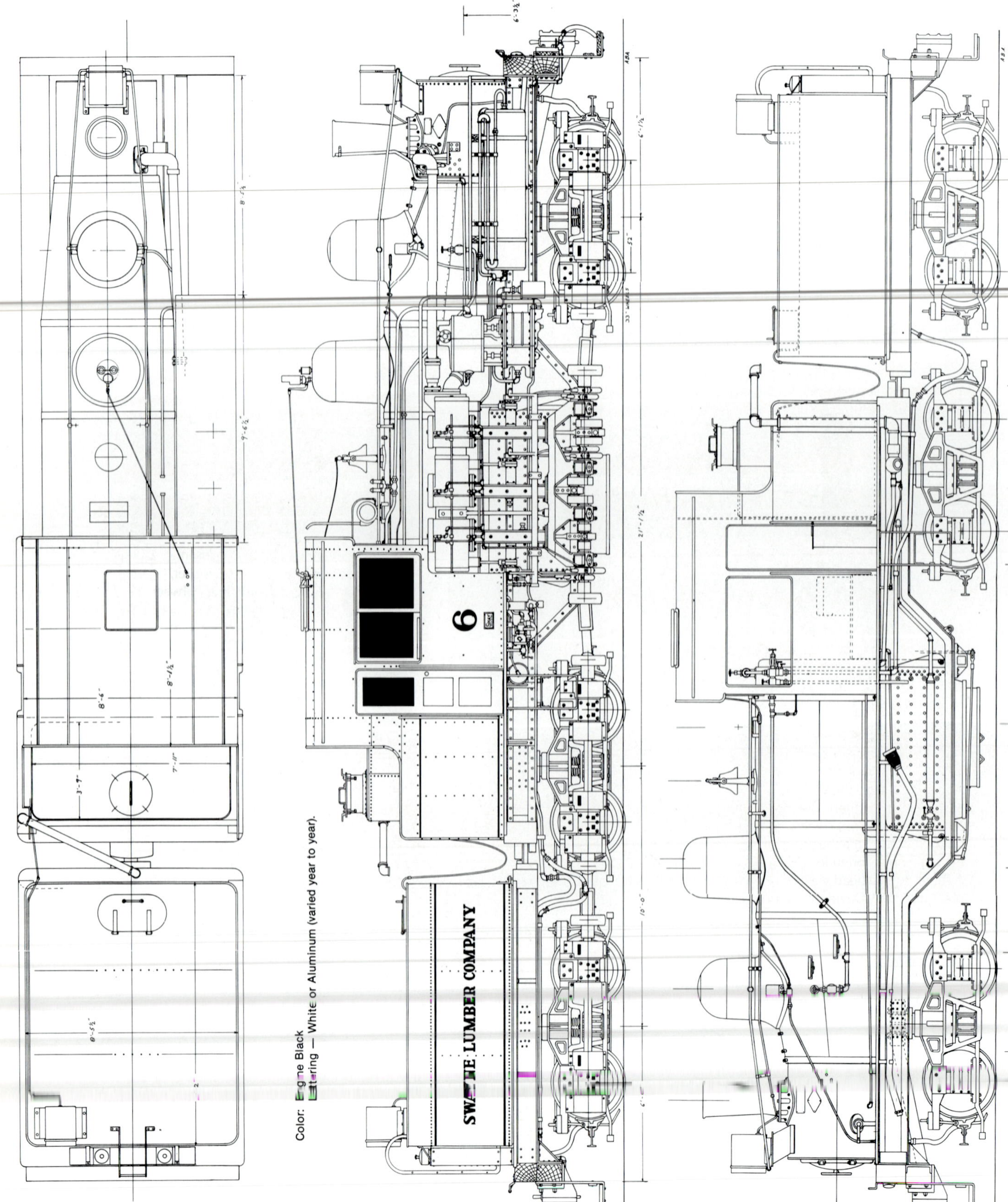

Scale: 3/16 inch = 1 foot (Full size Sn3 scale)

Above: No. 6 on the millyard wye in 1938. Note the Sacramento Northern overhead wire just above the tender. (Ted Wurm) Below: A rare lefthand view by Otto Perry showing the siphon, siphon hose, fire fighting reel and wooden lockers at the rear of the tender. At right: A backhead view of No. 6, taken at the Lima plant in 1926. (David W. Braun collection)

Scale: 3/16 inch = 1 foot (Full size Sn3 scale)

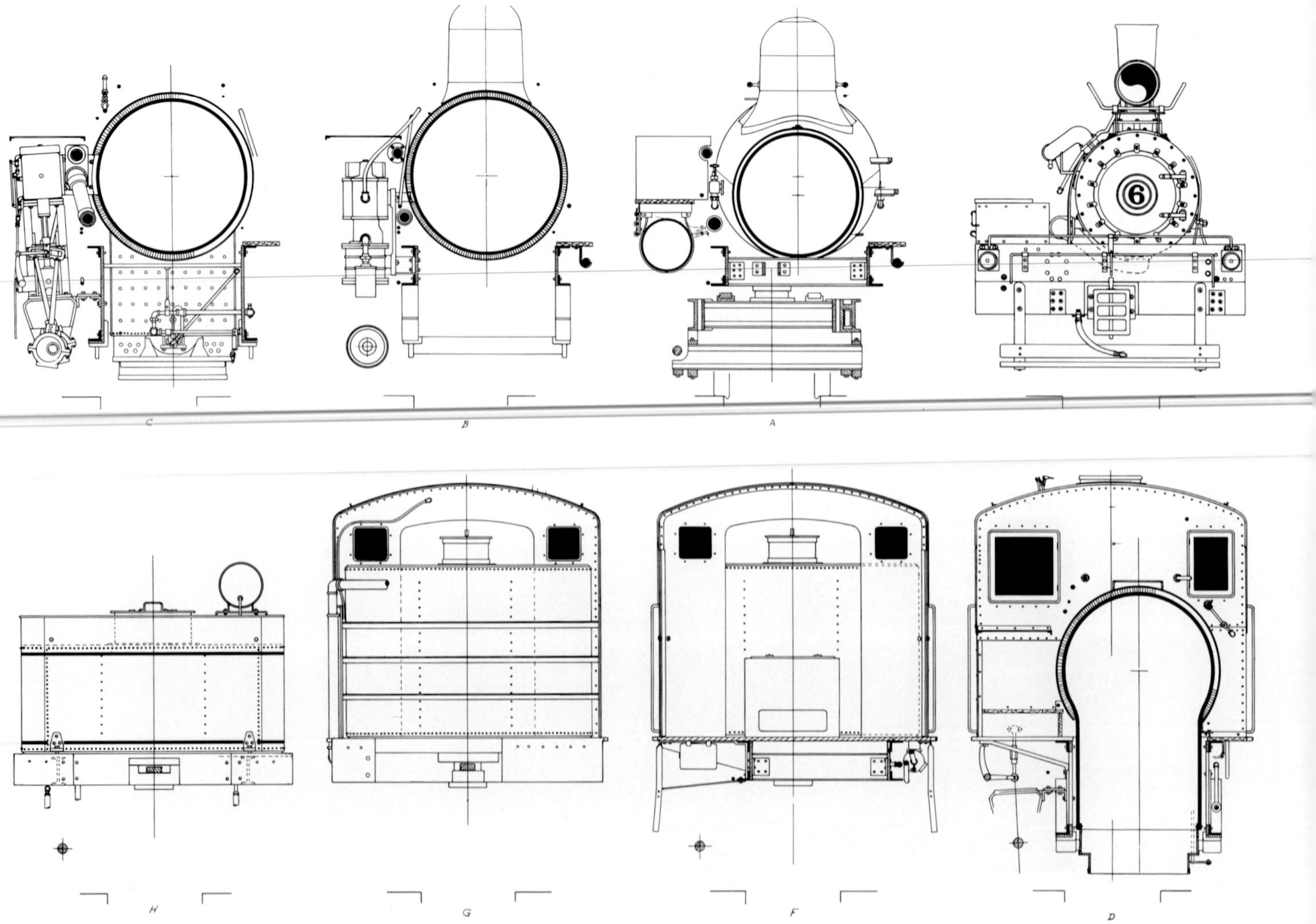

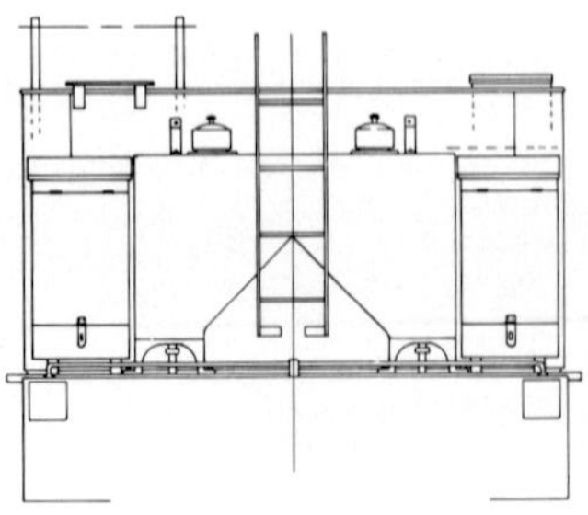

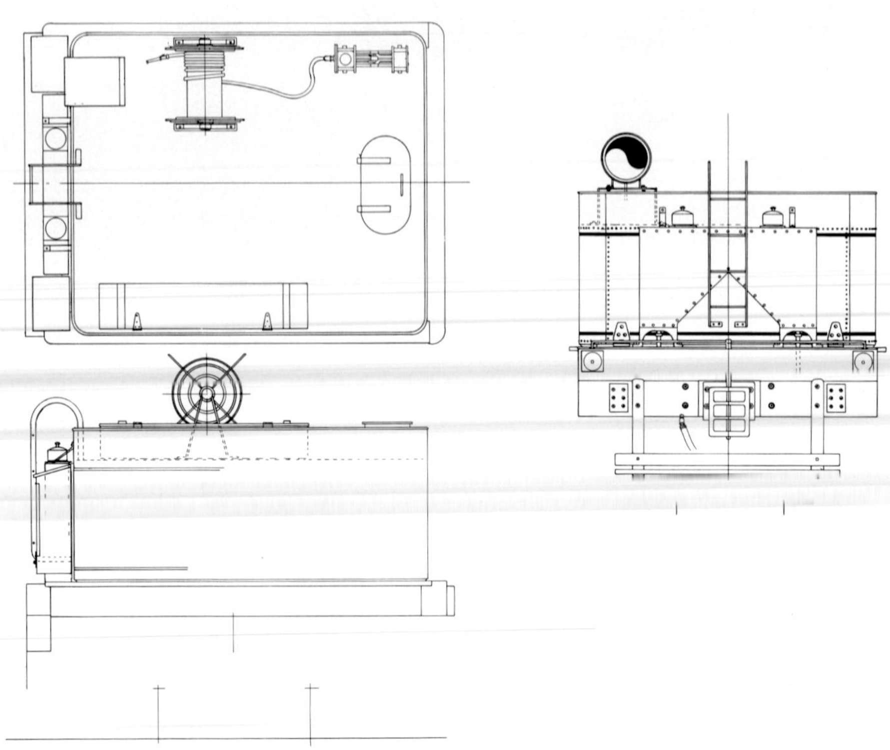

Swayne No. 6 Modifications.

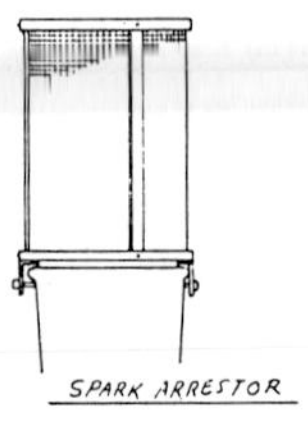

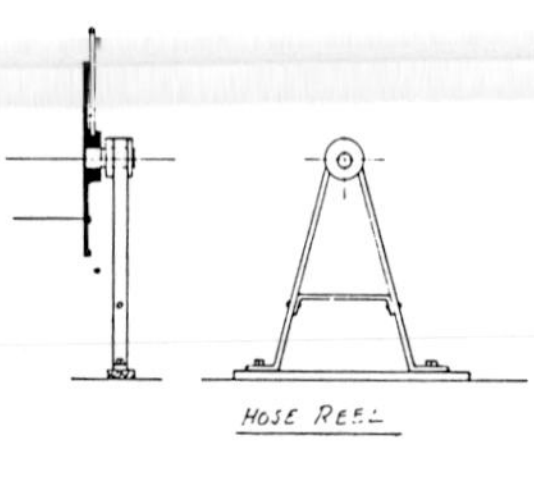

At left: Ex-Swayne No. 6, now West Side No. 12, at Tuolumne in 1944 and 1948, repectively. (Both photos John Cummings; Bert Ward collection)

West Side No. 12, ex-Swayne No. 6, is shown on the tourist West Side & Cherry Valley Railway. (Glenn Tyra; Paul Beckstrom collection)

Butte & Plumas No. 4, now Roaring Camp & Big Trees No. 7, still going strong in 1988, hauling tourists in Felton, California. (Paul Beckstrom)

THE THREE SISTER SHAYS

Swayne Number Six Shay and her two "sisters" formed a unique group providing an interesting "by-way" in Western logging and Shay locomotive lore. These Shays, while all built from the same erecting card, reflected the individual differences specified by the three customers involved: Madera Sugar Pine, New Mexico Lumber and, of course, Swayne. These locomotives make an interesting study, individually and as a group.

There was a total of eight Shays constructed to the general arrangement of Card 428-C-50-55 with three built to 36-inch narrow gauge. The Swayne Number Six was the first N.G. locomotive (third engine) built of this design, under Order 462. The second narrow gauge locomotive (fourth engine) was for the Madera Sugar Pine Company, Order 466, and the last narrow gauge engine (seventh locomotive) was Order 500 for the New Mexico Lumber Company.

The Swayne engine was ordered through California dealer N. B. Livermore and was shipped April 21, 1926. As built, it was equipped with D-valves, a superheater, steel cab and cast-steel trucks. The engine was shipped on a flatcar and, to save on shipping weight, Swayne specified the omission of the firebrick, oil pan and the steam brake equipment, intending to install this equipment themselves. The cost to Swayne, FOB Lima, Ohio, was $21,735.00.

The Madera Sugar Pine Shay was also ordered through Livermore on November 24, 1926, but was not shipped from Lima until February 5, 1927. The differences between this locomotive and Swayne's were the fuel supply, stack, air compressors and superheated injectors. But, like Swayne No. 6, it was also superheated.

The most important difference, however, was that the Madera Sugar Pine Shay came equipped with piston valves using the Baker Marine kit discussed in the text. Madera also specified flange oilers and a change in the gear ratio for which Lima charged $213.00. The engine was shipped on a flatcar to Madera, California where it was hauled by team-

—Photographs from the Collection of Allen County Historical Society, Lima, Ohio

sters to the company's mill, as Madera Lumber had no outside rail connections. The FOB cost was $22,357.00. The locomotive later worked at Delleker, California for the Feather River Lumber Company before going to the Michigan-California Lumber Company in 1944. Assigned (3rd) No. 6, the engine was used as a road locomotive from Camps 14 and 15 for the haul to the mill at Pino Grande. The engine was finally cut up for scrap in 1951.

The New Mexico Lumber Company engine was ordered through Seattle dealer, Hofius Steel, and shipped November 20, 1929 on three standard gauge shipping trucks rather than sending it out on a flatcar. The FOB cost to the New Mexico Lumber Company was $23,025.00. This Shay was equipped as a coal burner and featured Priest-type flanger blades on the first and third trucks and special snow covers on the gears for operation in the harsh Colorado winters. The engine was hauled to Montrose where it was placed on narrow gauge trucks and taken by the D&RGW to Ridgway. From there it was turned over to the Rio Grande Southern which towed it to the NMLC/RGS connection near Dolores. The Shay was then taken to the NMLC mill at McPhee, Colorado, located west of Dolores. Due to the fact that the New Mexico Lumber Company was shut down in January 1930, shortly after the locomotive was delivered, it is doubtful that it saw much logging service. Accordingly it was repossessed by Hofius, and on March 26, 1931 it was again hauled out by the Rio Grande Southern, this time in the other direction, south to Durango. The Shay was towed to Alamosa by the D&RGW where it was loaded onto a flatcar consigned to Hofius in Seattle. It was not sold again until 1936, when it became Oregon Lumber Company No. 7, operating in the Sumpter Valley region. The locomotive was converted to oil by the Sumpter Valley Railway at their Baker (Oregon) shops. NMLC/OLC No. 7 still exists, although in severely damaged condition.

Photographs of New Mexico Lumber Company No. 7 in service are unknown. Almost as rare are pictures of the Shay in service for its second owner, The Oregon Lumber Company. Photographer Harold Vollrath found the other Swayne "sister" locomotive at Bates, Oregon in 1946, after the railroad had shut down. The only further use for No. 7 would be on the scrap trains in 1947. Oregon Lumber changed the stack and removed the flanger equipment. The engine remained a wood-burner from 1936 until 1941 when it was converted to oil in the Sumpter Valley shops. In this view, the special snow covers, ordered by NMLC, remain in place. In a photograph by Al Phelps, ex-Madera Sugar Pine "sister" No. 6 is near Michigan-California's mill at Pino Grande on October 26, 1945, working the log dump. This was not its usual job, since it was the road engine for the haul to and from Camp 15. Michigan-California cut the engine up for scrap in 1951.

Sources

CHAPTER ONE

1. Knowles, C. D., A History of Lumbering in the Truckee Basin, 1942, published by the U.S. Forest Service, Nevada City, California.
2. Truckee Republican, December 2, 1874.
3. Tahoe World, October 8, 1980.
4. The Marine Digest, September 13, 1941.
5. Truckee Republican, October 21, 1890.
6. Myrick, David F., Railroads of Nevada and Eastern California, Vol. 2, 1963, page 897.
7. Reno Gazette, December 13, 1883.
8. Knowles, ibid.
9. Knowles, ibid.
10. Transcript, California State Railroad Commission hearing, April 12, 1913; all three quotes cited.
11. Truckee Republican, June 15, 1901.
12. Myrick, ibid, Vol. 1, page 431.
13. Railroad Commission, ibid.
14. Nevada County Recorder, Book 96, page 154.
15. Nevada Daily Transcript, June 9, 1893.
16. Truckee Republican, 1893.
17. Placer County Recorder, Auburn, California.
18. Placer County Recorder, Auburn, California.
19. Truckee Republican, November 6, 1909.
20. Indenture, Truckee Lumber Company to H. L. Lanfar, December 16, 1909.
21. Landfar Document Collection, Dana E. Supernowicz.
22. Truckee Republican, January 12, 1910.
23. Truckee Republican, May 9, 1910.
24. Truckee Republican, June 13, 1910.
25. Railroad Commission, ibid.
26. Landfar Documents, ibid.:
    —United States Department of Agriculture letters to Sierra Pacific Power Company, August 4, 29 and December 12, 1944; ditto, letter to Southern Pacific Company, November 30, 1944;
    —Letter of August 12, 1944 from West Side Lumber Company to Regional Forester, San Francisco.
    —Internal Memorandum, October 13, 1969, Sierra Pacific Power Company, discussing deed situation regards West Side Lumber, H. L. Lanfar, Kruger Estate Company, Truckee Lumber Company and Haslett Estate.
27. Truckee Republican, October 30 and December 25, 1909.
28. Plumas County Court House Incorporation Records, #599.
29. Signor, John R., Rails in the Shadow of Mt. Shasta, 1982.
30. Bornefeld, Ethyl L., A Report of Activity in the French Creek and Chino Creek Areas, 1890-1938, 1976, published by the U.S. Forest Service, Oroville, California.
31. American Lumberman, July 1906.
32. Bornefeld, ibid.
33. San Francisco Call, May 16, 1909.
34. Oroville Daily Register, September 22, 1909.
35. Pioneer Western Lumberman, January 1910.
36. Oroville Daily Register, June 6, 1910.
37. Map of Butte County, 1930, M. E. Polk, C.E.
38. Bornefeld, ibid.
39. Bornefeld, ibid.
40. Oroville Daily Register, June 9, 1911.
41. Oroville Daily Register, September 15, 1911.
42. Oroville Daily Register, September 28, 1911.
43. Truckee Republican, August 22, 1912.
44. West Side Collection, Russ Simpson, 1986.
45. Lanfar Documents, ibid., USDA letter of August 4, 1944.
46. Krieg, Allan, Last of the Three Foot Loggers, 1962.

CHAPTER TWO

1. Crocker, H. S., Crocker-Langley, San Francisco Directory, 1909.
2. Stindt, Fred A., The Northwestern Pacific Railroad, 1964.
3. The Marine Digest, April 17 and May 22, 1943.
4. Gardner, R. R. Co., Report of Timber Holdings of Truckee Lumber Company and Swayne Lumber Company, 1922, Russ Simpson Collection.
5. Butte County Deeds, Book 111, page 361.
6. Bornefeld, Ethyl L., Camp Enjoyment: The First Resort Along the Feather River, Butte County Historical Society Diggins, December 1976, page 87-98.
7. Bornefeld, A Report of Activity, etc., ibid.
8. Pacific Coast Wood and Iron, November 1, 1909.
9. Oroville Daily Register, November 16, 1909.
10. Oroville Daily Register, August 29, 1910.
11. Gardner, R. R. Co., ibid.
12. Bornefeld, Ethyl L., A Report of Activity, ibid.
13. Oroville Daily Register, July 25, 1911.
14. Oroville Daily Register, May 27, 1911.
15. Oroville Daily Register, July 25, 1911.
16. Bornefeld, Ethyl L., A Report of Activity, ibid.
17. American Lumberman, August 23, 1913.
18. Butte County Deeds, Book 137:415, pages 419-420.
19. Oroville Daily Register, October 26, 1916.

CHAPTER THREE

1. West Side Collection ibid.
2. Butte County Official Records, Book 233, page 71.
3. Oroville Daily Register, April 11, 1917.
4. Wagner, Jack R., Gold Mines of California, 1970.
5. The Timberman, June 1917.
6. Oroville Daily Register, April 25, 1917.
7. Oroville Daily Register, April 10, 1918.
8. The Timberman, June 18, 1918.
9. Butte County Deeds, Book 164, page 211-12; Book 168, page 397.
10. Butte County Deeds, Book 171, page 368-70.
11. Dennison, Harry, interview, 1988.
12. Thomas, Ed, Oroville Mercury-Register, June 18, 1987.
13. Shoup, L. H., Loggers, High Leads, Donkeys and Shays: Railroad Logging on the Oroville District, Plumas National Forest, 1909-1938, 1985.
14. Thomas, Ed, Oroville Mercury-Register, June 18, 1987.
15. The Timberman, September 1923.
16. Thomas, Ed, ibid, August 7, 1987.
17. Thomas, Ed, ibid, August 7, 1987.
18. Thomas, Ed, ibid, August 7, 1987.
19. Slaughter, E. E. "Butch," interviews, 1987-90.
20. Thomas, Ed, ibid, August 7, 1987.
21. Dennison, Harry, interview, 1988.
22. Thomas, Ed, letter to authors, October 8, 1987.
23. Thomas, Ed, Oroville Mercury-Register, August 7, 1987.
24. Dennison, ibid.
25. Slaughter, ibid.
26. Dennison, ibid.
27. Slaughter, ibid.
28. Slaughter, ibid.
29. Dennison, ibid.
30. Slaughter, ibid.
31. Slaughter, ibid.
32. Thomas, Ed. ibid, August 7, 1987.
33. Dennison, ibid.
34. Dennison, ibid.
35. Dennison, ibid.
36. Slaughter, ibid.
37. Ferguson, E. R., interview with the former Diamond & Caldor Railway train dispatcher, by the authors, 1987.
38. Thomas, Ed. ibid, August 7, 1987.
39. Slaughter, ibid.

CHAPTER FOUR

1. The Timberman, January 1929.
2. The Timberman, July 1931.
3. Thomas, Ed, ibid, April 21, 1987.
4. The Timberman, August, 1936.
5. Thomas, Ed, ibid, April 21, 1987.
6. Thomas, Ed, ibid, June 17, 1987.
7. Shoup, ibid.
8. Brown, Nelson, C., Logging: The Principles and Methods of Harvesting Timber in the United States and Canada, 1934, 1949.
9. The Timberman, November 1929.
10. The Timberman, November 1931.
11. The Timberman, June, 1933.
12. The Timberman, November 1938.
13. Shoup, ibid.
14. Shoup, ibid.
15. Shoup, ibid.
16. Shoup, ibid.
17. Thomas, Ed., interview, 1987.
18. The Timberman, January 1935.
19. Sullivan, Ken, interview, February 21, 1991.
20. Thomas, Ed, Oroville Mercury-Register, June 18, 1987.
21. The Timberman, September 1936.
22. The Timberman, November 1938.
23. Shoup, ibid.
24. Bornefeld, Ethyl L., A Report of Activity, ibid.
25. Bornefeld, ibid.
26. Thomas, Ed, ibid, June 17, 1987.
27. Thomas, Ed, ibid, April 21, 1987.
28. Shoup, ibid.
29. Shoup, ibid.
30. Butte County Official Records, Book 233:71.
31. Thomas, Ed, ibid, June 18, 1987.
32. Thomas, Ed, ibid, June 18, 1987.

CHAPTER FIVE

1. Shoup, ibid.
2. Sturgeon, Walter, interview, 1984.
3. Brown, ibid.
4. Dennison, ibid.
5. Shoup, ibid.
6. Thomas, Ed, ibid, June 18, 1987.
7. Sturgeon, ibid.
8. Shoup, ibid.
9. Shoup, ibid.
10. Dennison, ibid.
11. Slaughter, ibid.
12. Slaughter, ibid.
13. Shoup, ibid.
14. Shoup, ibid.
15. Shoup, ibid.
16. Shoup, ibid.
17. Shoup, ibid.
18. Shoup, ibid.
19. Slaughter, ibid.

PICKERING No. 7

Although Pickering Lumber Company controlled West Side during much of the 1920's and first half of the 1930's, views of West Side equipment lettered for Pickering are somewhat rare; doubly so for any equipment from the Oroville area.

Here, Pickering No. 7 (nee West Side No. 7 and Butte & Plumas No. 4) is seen at Tuolumne in August 1934 during the Depression shutdown. The stack is covered, the front coupler is missing and the cab window awning has rotted away. Although the lettering on the tank indicates that Pickering is the nominal owner, West Side had taken over, again, a month previous to the date of this photograph.

D. W. Thickens; Larry Harrison collection

INDEX

— Bold Numbers Indicate Photos —

This war surplus Curtiss JN-4D "Jenny," photographed at either Chico or Oroville in 1931, was a prototype for future fire fighting air tankers. The Jensen Type I takes its name from Charles "Red" Jensen, an early Sacramento flyer who is generally credited as being the father of California's aircraft fire fighting program.

This Sitka Spruce, cloth-covered aircraft carried sixty gallons of water in two thirty-gallon hoppers located on either side of the cockpit. The plane was powered by a twelve-cylinder "Liberty" engine, manufactured by the Lincoln Motor Company, which later went on to build Lincoln automobiles.

Apparently Swayne Lumber Company, because of its steam-powered railroad, constituted a significant fire hazard and was required by the U.S. Forest Service to have an air tanker stand by during summer months for quick suppression of railroad right-of-way fires. To this day, air tankers are extremely effective in fighting small, remotely located "initial attack" fires. Dropping water from the air can often prevent a small fire from spreading into a major conflagration long before ground personnel and equipment can reach the site.

—Photograph courtesy of Bill Jamieson (California Department of Forestry and Fire Protection)

J

K

L

M

N

O

P

Q

R

S

T

U

V

W

Y

Z

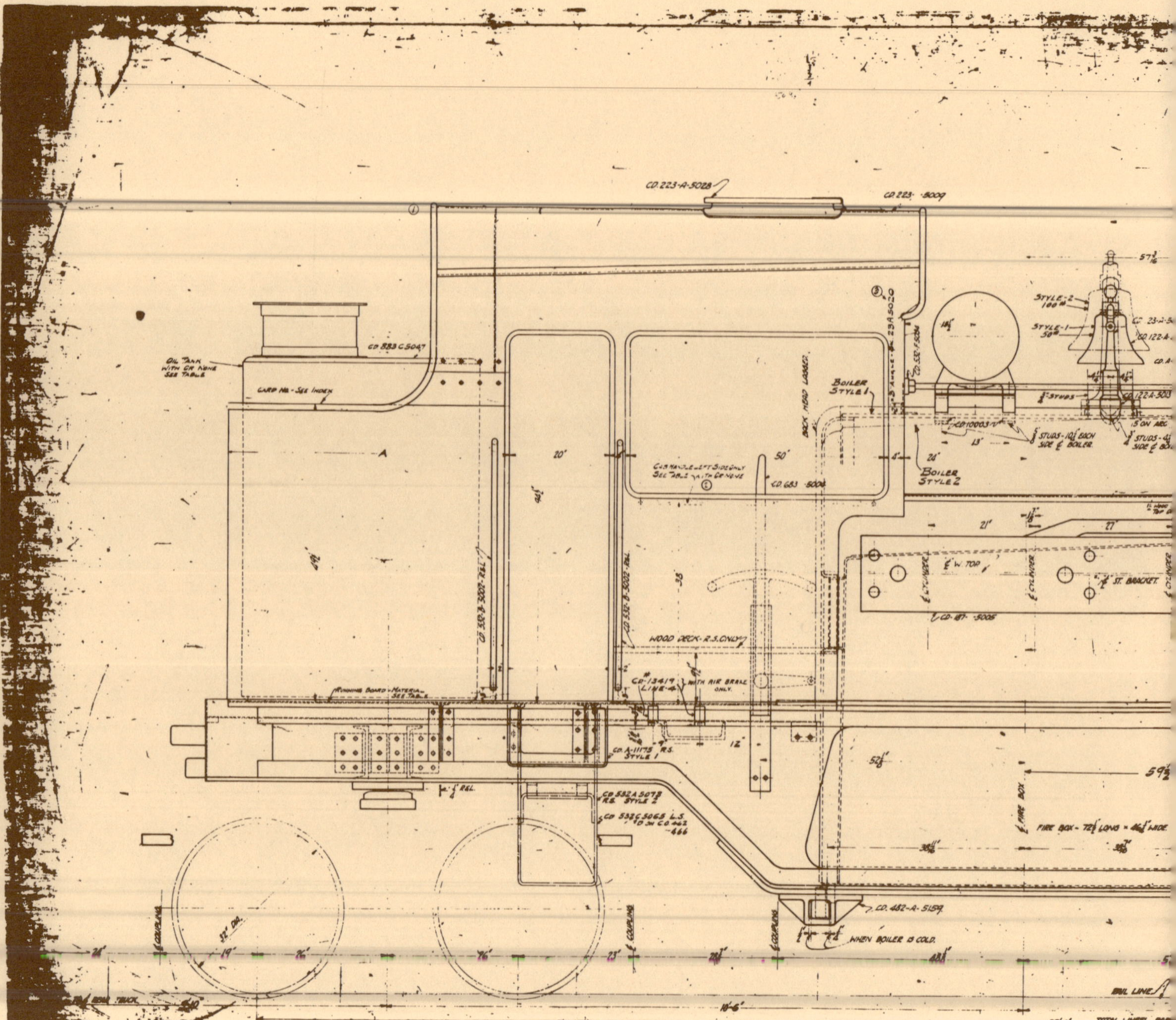
CD. 223-A-5028
OIL TANK WITH CR. NONE SEE TABLE
CARD NO. - SEE INDEX
BOILER STYLE 1
BOILER STYLE 2
BACK HEAD LAGGING
WOOD DECK - R.S. ONLY
WITH AIR BRAKE ONLY.
WHEN BOILER IS COLD.
CD. 482-A-5159
20"
50"
21"
27"
10'-6"
25'-6"